FPGA EDA

Kaihui Tu · Xifan Tang · Cunxi Yu ·
Lana Josipović · Zhufei Chu

FPGA EDA

Design Principles and Implementation

Springer

Kaihui Tu
Beijing, China

Xifan Tang
Los Gatos, CA, USA

Cunxi Yu
University of Maryland, College Park
College Park, MD, USA

Lana Josipović
ETH Zürich
Zürich, Switzerland

Zhufei Chu
Ningbo University
Ningbo, China

ISBN 978-981-99-7754-3   ISBN 978-981-99-7755-0 (eBook)
https://doi.org/10.1007/978-981-99-7755-0

This Springer imprint is published by the registered company Springer Nature Singapore Pte Ltd.
The registered company address is: 152 Beach Road, #21-01/04 Gateway East, Singapore 189721, Singapore

Paper in this product is recyclable.

Foreword

As the need for highly efficient systems has grown and the performance gains from Moore's Law have diminished, hardware acceleration has become ever more important. At the same time, chip design costs are skyrocketing, and application needs are rapidly changing, putting a completely custom silicon system out of reach for most applications. The unique, low-level hardware programmability of field-programmable gate arrays (FPGAs) has therefore become a key part of many systems, as a single chip can be programmed to enable a hardware solution for multiple applications.

To architect, implement, and program FPGAs requires sophisticated electronic design automation (EDA) tools, however, and indeed the greatest challenge of implementing a new FPGA architecture and making it usable has historically been the creation of these EDA tools. This book fills a crucial gap, as it covers all aspects of FPGA EDA tools. Some of these tools are similar to their custom (application-specific integrated circuit, or ASIC) cousins, but others are uniquely specialized to the implementation or targeting of FPGAs, and this book provides both the general overview and FPGA-specific discussion needed to fully understand FPGA EDA.

This work takes a holistic approach to FPGA EDA that is timely and important. Uniquely, it covers three sub-areas of FPGA EDA that together enable not only designing to and programming an FPGA, but also architecting a new FPGA and automating much of the implementation of its circuitry and layout. The first portion of this book explains the EDA tools needed describe a new FPGA architecture and evaluate its quality—crucial steps in making sure a new FPGA is usable and performant. Next it summarizes and compares a novel class of EDA tools that has emerged to rapidly implement the circuitry and layout of a new FPGA architecture, verify its correctness, and model its low-level characteristics. The higher implementation productivity enabled by these new tools is fueling wider adoption of embedded FPGA blocks within system-on-chip (SoC) designs and the development of application-tuned standalone FPGAs. Finally, this work details the full design and programming flow needed by end users of an FPGA, from high-level synthesis, through logic synthesis, physical implementation, and bitstream generation.

This book is a key reference not only for FPGA end users and EDA developers but also FPGA architects. Its coverage of both current industrial practice and emerging approaches makes it relevant to both FPGA practitioners and researchers. As FPGA designs become more complex, FPGA architectures proliferate, and embedded FPGAs find their way into more SoCs, knowledge of the full spectrum of FPGA EDA tools will be crucial to system architects and this book will be one of their key resources.

Toronto, Canada
August 2023

Vaughn Betz

Preface

The electronic design automation (EDA) for field programmable gate array (FPGA) is a special interdisciplinary study area. EDA software runs throughout the entire life cycle of FPGA design, including chip design stage and application design stage. With the ever-increasing demand for customized and application-specific electronic systems, FPGAs have become a popular choice for implementing complex digital designs and FPGA EDA is the key to make that happen.

This book aims to provide a comprehensive and up-to-date principles and implementations of FPGA EDA technology, absorbing nutrients from both industry and academia. It is written for students, researchers, and practicing engineers who are interested in understanding the design and optimization of FPGA EDA system.

The book begins with an introduction to two fundamental concept: FPGA and EDA. It then covers the databases and methodologies across the whole design flow: at chip design stage, EDA tools help chip design engineers to explore the FPGA architecture and complete their intricate masterpiece, while at application design stage, EDA tools help application design engineers to etch their brilliant thoughts into circuits.

Writing a book is a challenging and rewarding experience. When writing this book, we realized the vastness of the FPGA EDA field and the complexity of this topic. We tried to make the book as accessible as possible by explaining the concepts in a simple and intuitive way. However, there is always room for improvement, and we welcome any feedback or suggestions from the readers.

Writing a book is definitely not a solitary endeavor, authors are from across Asia, Europe, and North America, and all of them have invested their huge amount of time and effort to finish this course. Meanwhile, I am thankful to the people who have provided their precious assistance and support throughout the writing process: Dr. Lijiang Gao, Dr. Borui Cai, Prof. Rui Zhang, Dr. Grace Zgheib, Dr. Colin Yu Lin, Dr. Zhihong Huang, Dr. Yi Shu, Dr. Tianwen Li, Dr. Yuanming Zhu, Dr. Junying Huang, Yu Bao, Dong Zhang, Jian Han, Larisa Li, Yuan Li... We also would like to acknowledge the numerous researchers, scholars, and professionals whose work we have drawn upon in this book. Their contributions have been invaluable in shaping our ideas and perspectives.

Finally, I am grateful to the readers of this book, whose interest and support have made this project worthwhile. I hope that this book will prove to be a valuable resource for our audience and inspire them to further explore the fascinating world of FPGA EDA technology.

Beijing, China
April 2023

Kaihui "Kelvin" Tu

Contents

Part I
Introduction

Chapter 1
Introduction

Abstract Field programmable gate array (FPGA) is a typical semi-custom integrated circuit. The function of an FPGA is decided by both chip vendors and end users. Just like many other semiconductors, the design process of FPGA highly depends on Electronic Design Automation (EDA) tools. This chapter sorts out the complicated knowledge universe of FPGA technology, highlighting the EDA systems within. Due to FPGA's semi-custom characteristics, FPGA EDA is very distinctive and can be accordingly categorized into two different fields: chip design EDA and application design EDA.

1.1 FPGA Hardware Brief Introduction

1.1.1 FPGA Concept

FPGA is a semi-custom integrated circuit, that is, programmable units within it are pre-customized by vendors in the design house, after the device is manufactured and delivered to the end users, it can be "field"-customized for a second time to fully implement the desired functionality.

This characteristic makes FPGAs just like "lego" in the semiconductor world (Fig. 1.1), which means you can assemble (program) the pre-designed blocks (programmable units) into figures (circuit functions).

However, these "pre-design" and "assemble" tasks for FPGAs are too complicated to be carried out without computer's help; EDA is hereby introduced in to make them possible.

K. Tu et al., *FPGA EDA*, https://doi.org/10.1007/978-981-99-7755-0_1

Fig. 1.1 FPGA—The electronic "lego" of the semiconductor world

Fig. 1.2 World's first FPGA—Xilinx XC2064

1.1.2 FPGA Hardware Evolution

Xilinx Corporation (acquired by AMD in 2022) invented the first commercial FPGA–XC2064 (Fig. 1.2) in 1985. The device contains 64 programmable logic units consisting of two 3-input *Look-Up Tables* (LUTs) and an *Flip-Flop* (FF), enabling true "field" programmability for the first time. IEEE listed Xilinx XC2064 as one of the "25 Microchips that Shook the World" in 2009 [1] and inducted it into the "Chip Hall of Fame" in 2017 [2].

From that start point, FPGA has always stayed at the forefront of semiconductor technology and gradually forged a unique and complex knowledge system of itself, including FPGA chip design, FPGA chip design EDA, FPGA application design, FPGA application design EDA, FPGA foundry, FPGA assembly and testing, FPGA sales and marketing, etc. With more and more heterogeneous units are integrated into FPGAs, this knowledge system is expanding at an unprecedented speed. It is conceivable that if someone wants to run an FPGA business successfully, the first and most tricky thing to do is gathering a large group of professionals from a number of totally different fields. FPGA vendors, especially the top ones, have led the innovation for most of the time due to their strong ability to integrate supply chain resources from upstream to downstream and to maintain a diversed and interdisciplinary talent echelon.

FPGA hardware evolution can be inspected from several perspectives:

1. In terms of configuration memory
   Configuration memory that stores the configuration bit data is one of the fundamental components of FPGA. Just like the material of lego toys can be various, FPGAs accordingly can be categorized into several types in terms of configuration memory:

   a. SRAM (Static Random Access Memory) type
      Mainstream type since FPGA was born, the configuration data should be read into the on-chip SRAM when the power is turned on. After the configuration is completed, the configuration data in SRAM will be lost (volatile), and the internal logic function of the FPGA will also disappear. SRAM-based FPGAs are reusable and low-cost, but reloading may needs an external memory device's help.
   b. Antifuse type
      This type of FPGA can only be programmed once by burning the fuses within and was first introduced in 1992. Although loses the flexibility of reprogramming, it greatly improves the stability. FPGAs with this structure are more suitable for applications in harsh environments, such as high vibration and strong electromagnetic radiation and other aerospace fields. Because of the fixed logic, the device powers up instantaneously and consumes less power and size than other types of FPGAs.
   c. Flash type
      Flash memory is non-volatile, the FPGA of this type has the flexibility of SRAM structure and the reliability of anti-fuse structure at the same time. The cost of this technology is high, but the number of transistors used and leakage current are relatively small [3]. This type of devices is also very suitable for aerospace applications [4–6].
   d. Emerging technology
      To further advance non-volatile FPGA's PPA (Power, Performance and Area), emerging device technology is actively researched in the last two decades.
      Resistive Random Access Memory (RRAM) [7] and Magnetoeresistive Random Access Memory (MRAM) [8] are the two representative non-volatile

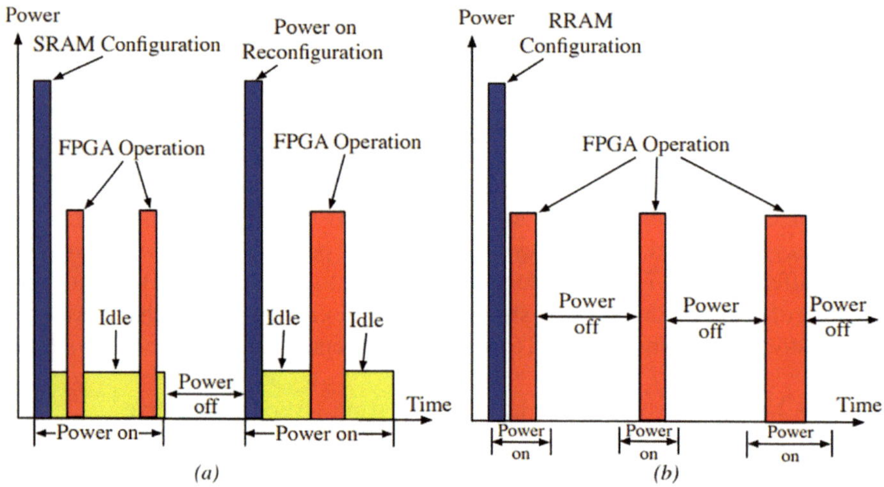

Fig. 1.3 Power consumption of **a** volatile (SRAM) FPGA and **b** non-volatile (RRAM) FPGA

memory devices, which have strong potential in substituting Flash memory in FPGAs (Fig. 1.3).

Both technologies share common advantages over Flash memory in

i. compatibility to BEoL (Back End of Line) process, leading to significantly higher area density. Unlike Flash transistors, RRAMs and MRAMs are fabricated between metal layers, no longer consume transistor area [9, 10].
ii. low read/write voltages as well as current, being similar to nominal voltages of logic transistors. This avoids dedicated circuits for accessing memory elements, which require different process than regular transistor and force PPA overhead [9, 11].
iii. fast read and write speed at the level of nanoseconds, which can reduce programming time and power of FPGA devices [12, 13].

Nowadays, FPGAs based on emerging technologies have attracted intensive research interests and are considered as next generation of FPGA technology (Table 1.1).

These FPGAs are suitable for:

i. IoT (Internet of Things) applications which require both ultra-lower-power and high-performance at the same time.
ii. aerospace applications which require highly robust against radiation-induced soft errors[14].

We refer interested readers to [15–19].

2. In terms of component resources

Modern FPGAs are composed of various "island-type" units and "ocean-like" interconnect resources among them. *Tile*, as the first-level sub-unit of FPGA,

Table 1.1 Comparison of FPGAs based on different programmable memory type

Memory type	Volatile	Programmability	Latency	Power	Area	Cost
SRAM	Yes	Repeatable	Low	Medium	Large	Low
Antifuse	Non	Once	Very low	Very low	Very small	Very high
Flash	Non	Repeatable	High	Low	Small	Medium
MRAM	Non	Repeatable	Low	High	Small	High
RRAM	Non	Repeatable	Low	High	Small	High

is composed of more fine-grained "*Site*" units. *Tiles* of the same type (excluding special types like I/O) are usually arranged in columns, and this layout can help approximate the distances between each general *tile* and I/O *tile*, so that there can be relatively high flexibility for physical implementation. Each *tile* has a unique address throughout the FPGA, which is typically represented by two-dimensional (for single-die) or even three-dimensional (for multi-die [20]) coordinates (Fig. 1.4).

The component resource types are rapidly enriched with the development of programmable technology. State-of-the-art FPGAs have even evolved into a hybrid device with multiple types of architectures. The current mainstream component resource types include:

a. Generic Logic Tile (Spatial Computing Tile)
   Generic Logic Tile (GLT) is the fundamental programmable unit distributed in the FPGA, and the site under this unit is called Generic Logic Site (GLS). GLS here is generally composed of LUT/FF, and the structure of the GLT/GLS can be various.
   AMD GLT—CLB [21]:
   Configurable Logic Block (CLB) is the main resource for AMD FPGA to implement basic sequential and combinational circuit functions (Table 1.2). Taking AMD's Versal architecture as an example, each CLB contains 4 GLSs, and each GLS contains 8 adaptive LUT6s and 16 FFs. AMD GLS has two types: SliceL (Logic) and SliceM (Memory). The latter enables the LUT under the GLS to be used as distributed memory by adding independent write addresses, write enable and clock signals. By enabling distributed memory, the maximum capacity of the chip memory is increased, and the memory usage efficiency is improved.
   Intel GLT—LAB/MLAB [22]:
   Logic Array Block (LAB) is the main resource for Intel FPGA to implement basic sequential and combinational circuit functions (Table 1.3). MLAB (Memory LAB) is a superset of LAB. In addition to all LAB functions, it supports dual-port SRAM up to 640 bit as a distributed memory. Taking Intel's Hyperflex architecture as an example, each LAB/MLAB contains 10 GLSs, and each GLS contains 1 adaptive LUT6 and 4 FFs. Intel GLS is called Adaptive Logic Module (ALM), and its structure is slightly different in LAB and MLAB.

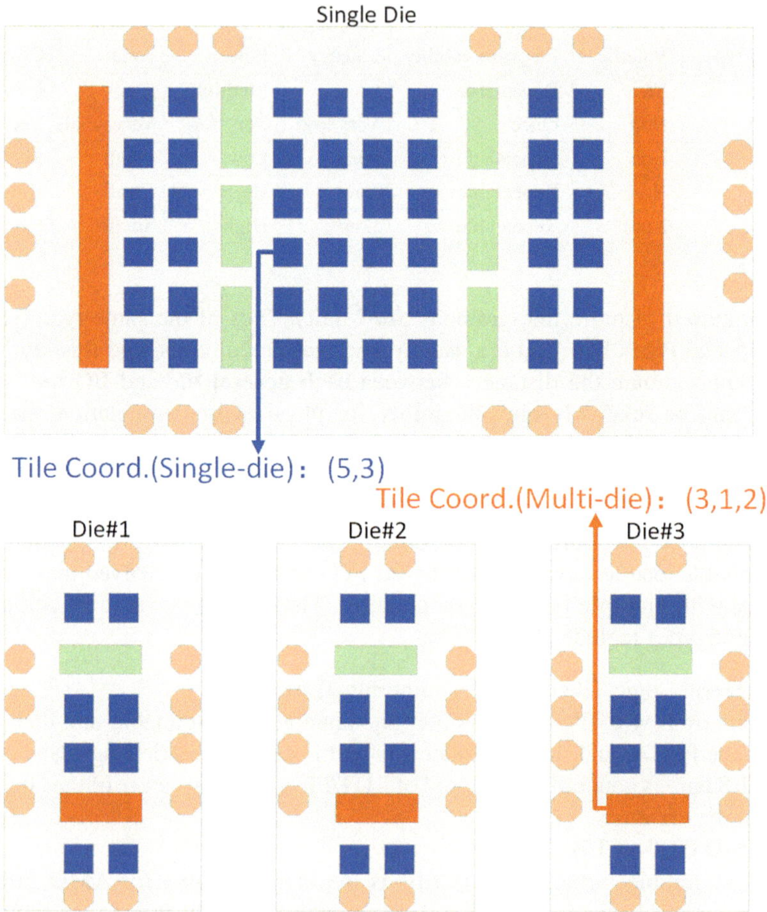

Fig. 1.4 Tile coordinate systems in single-die/multi-die FPGA

Table 1.2 AMD GLT resources—Versal architecture

AMD GLT	AMD GLS	LUT6	FF	Carry chain	Distributed RAM	Shift register
CLB	4 (2 SliceL + 2 SliceM)	32	64	4	2048 bits	1024 bits

Table 1.3 Intel GLT resources—HyperFlex architecture

Intel GLT	Intel GLS	LUT6	FF	Carry chain	Distributed RAM	Shift register
LAB	10 (ALM)	10	40	1	/	/
MLAB	10 (ALM)	10	40	1	2048 bits	1024 bits

b. Input/Output Tile
Input/Output Tile (IOT), powered by groups of banks, is the interface between FPGA and the outside world. There are many types of IO in modern FPGAs, categorized into single-ended IO (LVTTL, LVCMOS, DDR, etc.) and differential IO (LVDS, LVPECL, SerDes, etc.). Single-ended signaling is the simplest and most commonly used method for transmitting electrical signals between devices. The signal is represented by a varying voltage on one wire; however, the dynamic power consumption of single-ended IO increases exponentially with the increase of clock frequency, so it is not suitable for application in high-speed circuits. Then comes the differential IO, it uses two wires for each signal (differential pair), which has better noise immunity than single-ended IO.

c. Clock Management Tile
Clock Management Tile (CMT) is a firmware resource dedicated to clock synthesis, elimination of clock skew, and clock phase and frequency adjustment in the FPGA. By programming, high precision, low jitter frequency multiplication, frequency division, duty cycle adjustment, and phase shift of the clock can be achieved. Delay Locked Loop (DLL) and Phase Locked Loop (PLL) are two common CMT implementations. DLL is based on the digital sampling method, which inserts a delay between the input clock and the feedback clock so that the rising edges of the input clock and the feedback clock are consistent. It is also called a digital phase-locked loop. The PLL, also known as an analog phase-locked loop, uses voltage to control the delay, and uses a voltage-controlled oscillator (VCO) to achieve a delay function similar to that in the DLL, also known as an analog phase-locked loop.

d. Memory
There are different types of memory for FPGAs (Fig. 1.5) listed below.
On-chip memory (Memory Tile, MMT):
On-chip memory means FPGA integrates the memory tile as a hard core. Block RAM (BRAM) is a typical representative of traditional on-chip memory. This tile can be programmed as single port, simple dual port , true dual port, read-only memory or other modes, and the depth and width of the stored data can also be configured with high flexibility. The Distributed RAM configured by the look-up table is an effective supplement to the BRAM and is suitable for

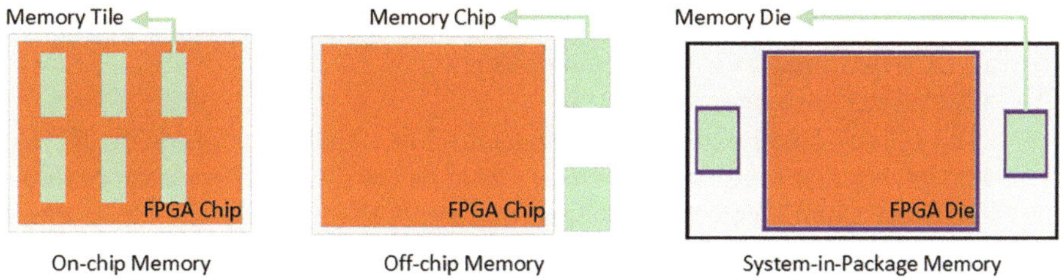

Fig. 1.5 Different types of memory for FPGAs

small storage requirements. The location of the BRAM resources in the FPGA is fixed and generally distributed in columns, which may cause a long wiring delay to the general logic, the use of distributed storage can help to ease the pain.

Off-chip memory (Memory Chip):

Off-chip memory is a memory that places on the periphery of an FPGA and provides extended storage space for it. A typical representative is DDR RAM.

System-in-Package memory (Memory Die):

System-in-Package (SiP) memory is a memory that integrated with general logic through an interposer to further expanding storage space. The typical representative is High Bandwidth Memory (HBM), which compactly connects stacked Dynamic Random Access Memory (DRAM) and FPGA through an interposer. This technology makes it possible for AI algorithms to be completely run on-chip, at the same time as the integration level is improved, the bandwidth is no longer limited by the number of interconnections of chip pins, so that the IO bottleneck is solved to a certain extent.

e. Scalar Computing Tile

SCT (Scalar Computing Tile), typically represented by Central Processing Unit (CPU) and Micro Controller Unit (MCU), has their unique advantages that traditional FPGA cannot achieve. Traditional FPGAs are good at parallel processing, and scalar engines are good at running control, the combination of the two can achieve a higher performance/watt ratio. Embedding SCTs in FPGA has become a common practice (e.g., ARM in AMD Zynq family/RISC-V in Microchip PolarFire family). Apart from such hard core implementations, some scalar engines are designed as soft cores (e.g., Intel's NIOS/AMD's MicroBlaze).

f. Vector Computing Tile

VCT (Vector Computing Tile), typically represented by Digital Signal Processing (DSP) unit and Graphics Processing Unit (GPU), is more efficient on processing a narrower set of parallel computing functions; however, it suffers from limited latency and efficiency due to inflexibility of memory hierarchies.

g. Matrix Computing Tile

Matrix Computing Tile (MCT) are purpose-built tiles that offer dramatic leaps in performance for AI workloads (matrix multiplication).

h. Other Analog/ASIC Tiles

Except for the tiles above, in modern FPGAs, many analog or application specific units are integrated, such as analog-to-digital/digital-to-analog conversion (ADC/DAC) units, video codec units, etc.

i. Clock Tree Resources

CTR (Clock Tree Resources) is a special set of signal paths inside the FPGA, like its vascular network. The clock signal on the clock network can ensure a relatively small signal skew, that is, make sure the time that the clock signal reaches each flip-flop is as close as possible.

j. Interconnect Resources

ICR (Interconnect Resources) provide communication channels that connect

Fig. 1.6 World's first SoC FPGA—Altera excalibur

all tiles together inside the FPGA, like its neural network. Traditional interconnect resources are controlled by programmable switches, which allow signals to switch to different paths. Modern FPGAs also use Network on Chip (NoC) interconnects that act like highways to accelerate the inner data transfer process.

Since the first heterogeneous computing engines (ARM SCT) was integrated into FPGA by Altera (acquired by Intel in 2015) in 2000 (Fig. 1.6), the era of SoC FPGA has begun. More and more types of architectures have been successfully coupled under the same umbrella (Fig. 1.7), and in [23], FPGA architecture progression has been well reviewed. Until 2019, Xilinx finally merges nearly all popular architectures in a single chip (Versal series) and pushing adaptive SoCs to a whole new level. Here is a brief history of FPGA resource evolution (Table 1.4).

3. In terms of process technology

   FPGA always pursuits the most advanced process node, since the beginning of 21st century, the big two (AMD/Xilinx and Intel/Altera) alternately lead the node shrinking race.

   a. Bulk/SOI

      Bulk is the traditional technology for FPGA. It is built on a standard silicon wafer. In contrast to bulk, Silicon-On-Insulator (SOI) makes use of SOI wafers, which incorporate a thin insulating layer within the substrate to suppress leakage. There are two types of SOI devices: partially depleted (PD-SOI) and fully depleted SOI (FD-SOI). Lattice (28 nm) and Quicklogic (22 nm) are on FD-SOI technology.

   b. Planar/FinFET/GAAFET

      Complementary metal-oxide semiconductors (CMOS) technology introduced planar transistors in the mid-20th century; however, the downsizing of planar transistors also brought numerous problems such as gate leakage currents, short channel effects, quantum tunneling leakage, variability, mobility degradation, etc. New technologies then come out to ease the pain, such as FinFET and GAAFET.

      FinFET (Fin Field Effect Transistor) technology is for relatively high-end

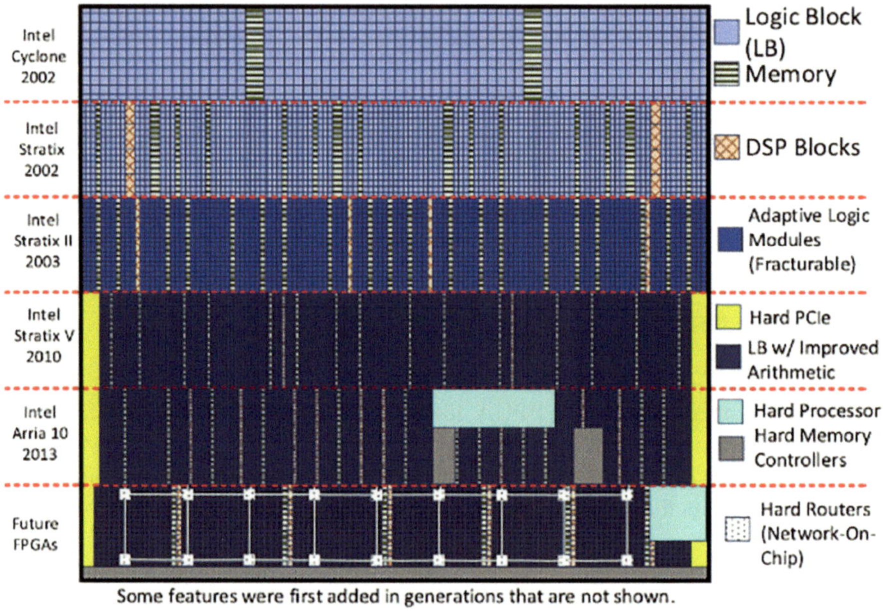

Fig. 1.7 Increasing heterogeneity of Intel/Altera FPGAs[24]

Table 1.4 FPGA resource-type evolution history

Resource type	First introduced family	Time (year)
GLT (LUT/FF)	Xilinx XC2000	1985
IOT (Basic)	Xilinx XC2000	1985
ICR (Basic)	Xilinx XC2000	1985
GLT (Carry chain/distributed RAM)	Xilinx XC4000	1991
MMT (BRAM)	Altera FLEX	1995
CMT (PLL)	Altera FLEX	1996
CMT (DLL)	Xilinx Virtex	1998
SCT (ARM)	Altera Excalibur	2000
VCT (Fixed point DSP)	Xilinx Virtex-II	2001
IOT (DDR IO)	Xilinx Virtex-II	2001
IOT (SerDes IO)	Xilinx Virtex-II	2001
SCT (PowerPC)	Xilinx Virtex-II Pro	2001
VCT (Float point DSP)	Intel Arria10	2015
VCT (GPU)	Xilinx Zynq Ultrascale+ MPSoC	2015
Other (RF)	Xilinx Zynq Ultrascale+ RFSoC	2017
MMT (HBM)	Intel Stratix10 MX	2017
ICR (NoC)	Achronix Speedster7t	2019
MCT (AI)	Xilinx Versal	2019
SCT (RISC-V)	Microchip PolarFire	2022

FPGAs (such as 16nm/14nm and beyond). It has a fin-shaped body – the silicon fin that forms the transistor's main body. FinFET devices display superior short-channel behavior and have considerably lower switching times, and higher current density than planar technology.
GAAFET (Gate-All-Around Field Effect Transistor) is a promising and futuristic transistor candidate to replace FinFET, since the channel width variations could cause undesirable variability and mobility loss as the fin width in a finFET approaches 5 nm (Fig. 1.8).
Just like planar technology, the source, gate, and drain can sit on an insulating layer (SOI) or on bare silicon (bulk). Most of the advanced Intel/AMD FPGAs (below 20 nm) are on bulk FinFET technology at the moment.

4. In terms of package technology

   a. Wire-bond/Flip-chip
   Wire-bond is the oldest and most common assembly technology. In a wire-bond FPGA, the IC is mounted to the substrate with the "active face" the face where the circuitry has been built-up. Small wires arch from the inputs and outputs ("I/Os") on the outside edges–known as the "periphery" of the IC to a specific on the substrate.
   Flip-chip has emerged as the best alternative to wire bond. The defining feature of the flip-chip package is a "flipped" IC, with the active side facing downward or toward the substrate. The interconnects in a flip-chip FPGA are much shorter than wire-bond, meaning that electrical losses and heat generation will be less severe.
   b. Planar/2.5D/3D
   Instead of traditional planar package solutions (single die), 2.5D package integrates multiple dies on a single interposer and interconnecting those chiplets on that interposer using metal interconnect. Intel/AMD's advanced FPGA devices commonly adopted this type of technology.
   True 3D package is a very advanced technology that splits FPGAs into multiple chips and stacks them. It still needs some time for true 3D FPGA to embrace mass production.

1.1.3 FPGA Compares With Other Architectures

1. A prototype platform for other architectures
The coolest Bugatti sports car in the world can be built with "lego" blocks (in fact, modern automobile manufacturing is also based on modularity), just like a cutting-edge architecture being prototyped by an FPGA. Using "lego" blocks to build a car that can actually drive will face many handicaps, for example, complex engine is difficult to implement, the solution would be installing a real engine (like

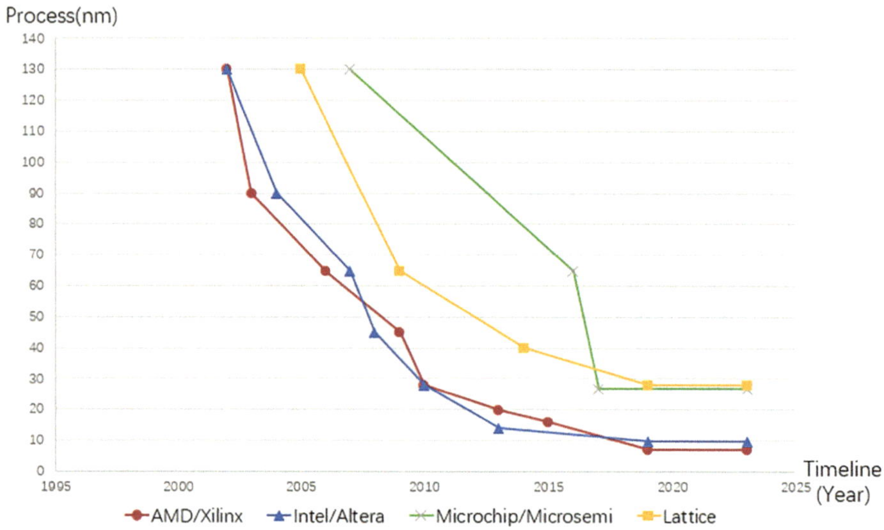

Fig. 1.8 FPGA giants' technology node cadence (as of 2023)

Fig. 1.9 "Lego" Bugatti Chiron and real Bugatti Chiron

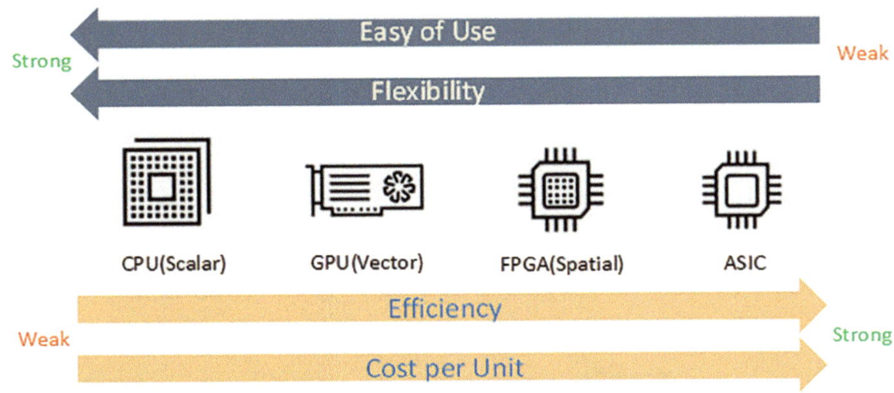

Fig. 1.10 FPGA compares with other computation architectures

embedding scalar computing tile) in it; tyres are also a problem, yet the simplest way is just put four real rubber tyres (like embedded analog circuit components) on it. This is another practice of heterogeneous integrating—put some "real car" parts in a "lego car".

Due to its hardware programmability, FPGA is an ideal prototype platform to simulate chip functions of any architectures in theory. When the shipment is not so huge (below around \$3m/400,000 pieces) or the function still needs iteration, FPGA, with zero *Non-Recurring Engineering* (NRE) cost and relatively high flexibility, is the choice (Fig. 1.9).

2. A computation platform with other architectures
   In terms of flexibility and easy of use, traditional FPGA has advantages over ASIC but cannot compete with software programmable scalar/vector architectures, however, if you look at performance and power efficiency or cost per unit, the story comes the opposite way. That's something called trade-offs (Fig. 1.10).

1.2 FPGA EDA Brief Introduction

1.2.1 FPGA EDA Concept

Electronic Design Automation, as its name suggests, is using computers to help electronic circuits design and therefore crowned as the "mother of chips". FPGA designs highly depend on EDA without exception.

The full design process of FPGA is divided into two stages: the chip design stage and the application design stage. The former is completed by the FPGA vendors, and

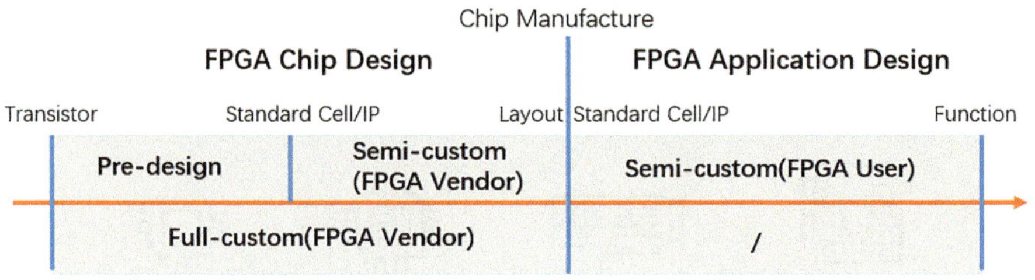

Fig. 1.11 FPGA design full flow: from vendors to users

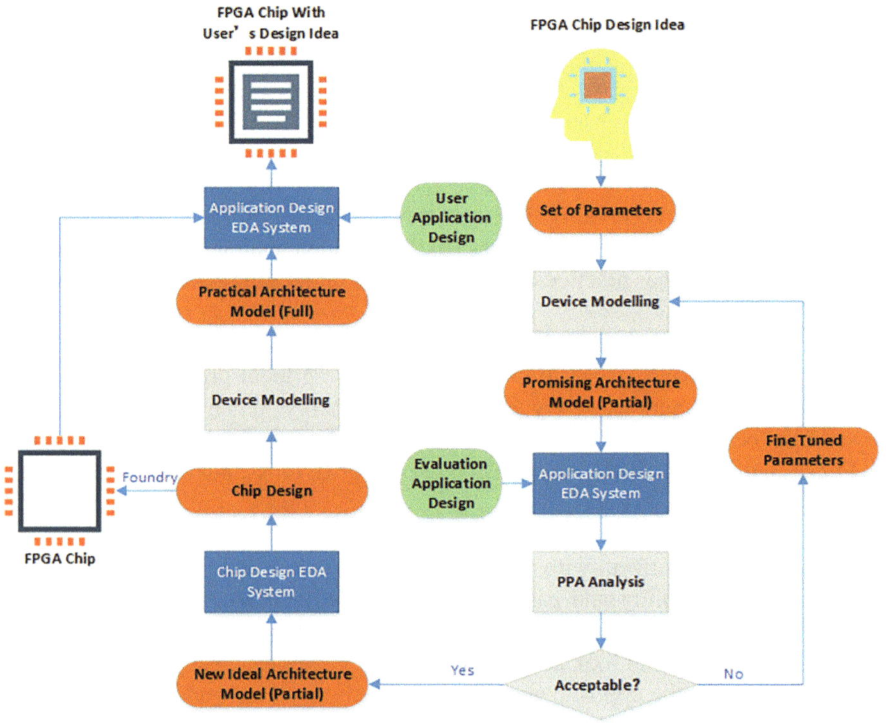

Fig. 1.12 FPGA EDA full flow: from vendors to users

the latter is handed over to the FPGA users. Both full-custom and semi-custom design methods are adopted during the chip design. Full-custom design requires designers to complete circuit design from the bottom-level, while another design approach, semi-custom design, uses pre-designed macros to simplify the effort (Fig. 1.11).

Correspondingly, FPGA EDA can also be divided into two parts: FPGA chip design EDA (to assist vendors in chip design stage) and FPGA application design EDA (to assist end users in application design stage) (Fig. 1.12). For FPGA chip design EDA, EDA vendors such as Cadence/Synopsys/Siemens provide univer-

Table 1.5 Different EDA systems for heterogeneous hardware architecture

Hardware architecture	Chip design EDA	Application design EDA
Spatial (Vanilla FPGA)	Virtuoso, Fusion compiler, etc.	Vivado, Quartus, etc.
Scalar	Virtuoso, Fusion compiler, etc.	Keil, IAR, etc.
Vector	Virtuoso, Fusion compiler, etc.	CUDA, etc.
Matrix	Virtuoso, Fusion compiler, etc.	TensorFlow Compiler, PyTorch JIT, etc.
ASIC	Virtuoso, Fusion compiler, etc.	/

sal design tools, and for FPGA application design EDA, FPGA vendors such as AMD/Intel/Microchip provide proprietary tools that only serve their own chips.

For modern SoC FPGAs, the integrated heterogeneous scalar/vector/matrix computing tiles have their own application design EDA systems different from vanilla FPGAs (Spatial). Although the industry is trying to unify them altogether (Intel oneAPI and AMD Vitis), vanilla FPGA EDA technology is currently still a self-contained system that cannot be replaced in a short period of time. The following chapters of this book mainly focus on vanilla FPGA EDA (Table 1.5).

1.2.2 FPGA Chip Design EDA

In FPGA chip design, full-custom method (designing based on transistors) is like "elaborately carving a painting on stones", so the performance of the chip can be maximized without wasting too much chip area; however, this design method consumes more labor and time; while semi-custom method (designing based on standard cells or IPs) runs through the classic synthesis-implementation flow and is much more quicker due to the higher level of automation, but sometimes cannot achieve the best result. Considering these trade-offs, only the integrated circuits that require extremely high performance, low power consumption and limited area will use the full-custom approach.

Figure 1.13 shows the brief flow of FPGA chip design. At the beginning of it, designers will evaluate the architecture based on the requirements and then do the module partition, deciding which modules should be designed with full-custom method and which modules should be designed with semi-custom method. After each module is finished on their own way, they are merged at the top level to form the final design.

Legacy EDA tools for FPGA chip design are abundant, and (Table 1.6) lists the most popular ones.

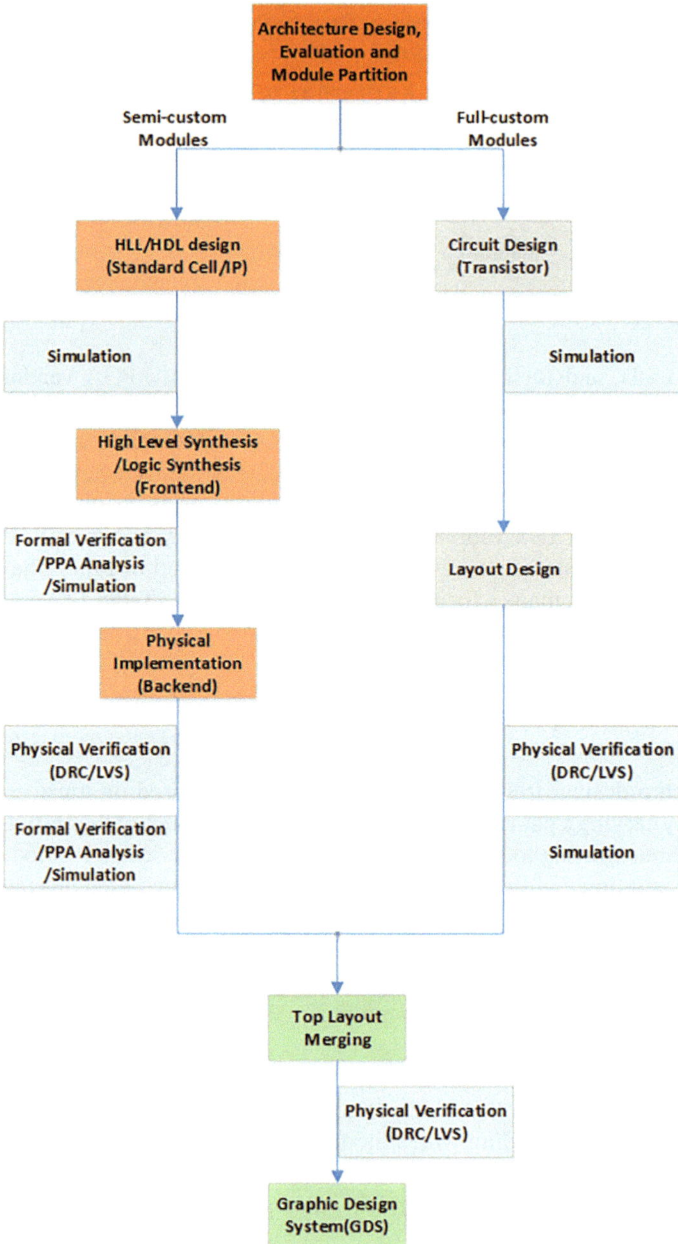

Fig. 1.13 Vanilla FPGA chip design EDA flow

Table 1.6 Iconic vanilla FPGA chip design EDA frameworks

Frameworks	Type	Maintainer	License	Description
Virtuoso	Full-custom	Cadence	Commercial	A holistic, system-based solution that provides functionality to do the custom IC design and sign-off
Fusion compiler	Semi-custom	Synopsys	Commercial	RTL-to-GDSII implementation system architected to address the complexities of advanced process node design
Calibre	/	Siemens EDA	Commercial	Well known IC sign-off verification and DFM (Design For Manufacturability) optimization tool
OpenFPGA	Semi-custom	CHIPS Alliance	MIT	The first open-source FPGA IP generator supporting highly customizable homogeneous FPGA architectures
COFFE	Full/Semi-custom	University of Toronto	MIT	An open-source FPGA design toolset for automatic FPGA circuit design
Australis	Semi-custom	QuickLogic	Commercial	Built on the OpenFPGA open-source framework that enables rapid prototyping of customizable FPGA architectures

Table 1.7 Iconic vanilla FPGA application design EDA frameworks

Frameworks	Maintainer	License	Description
VTR	University of Toronto	MIT	A world-wide collaborative effort to provide an open-source framework for conducting FPGA architecture and EDA research and development
F4PGA	CHIPS Alliance	Apache-2.0	A Verilog-to-Bitstream EDA flow targeting commercial FPGAs
RapidWright	AMD/Xilinx	Apache-2.0	Provide Vivado Interface for users to build customized FPGA implementations
Vivado	AMD/Xilinx	Commercial	A Classic FPGA EDA design suite that supports AMD/Xilinx devices
Quartus	Intel/Altera	Commercial	A Classic FPGA EDA design suite that supports Intel/Altera devices
Synplify	Synopsys	Commercial	Industry standard logic synthesis tool for producing high-performance, cost-effective FPGA designs
ModelSim	Siemens	Commercial	A multi-language environment for simulation of hardware description languages and includes a built-in C debugger

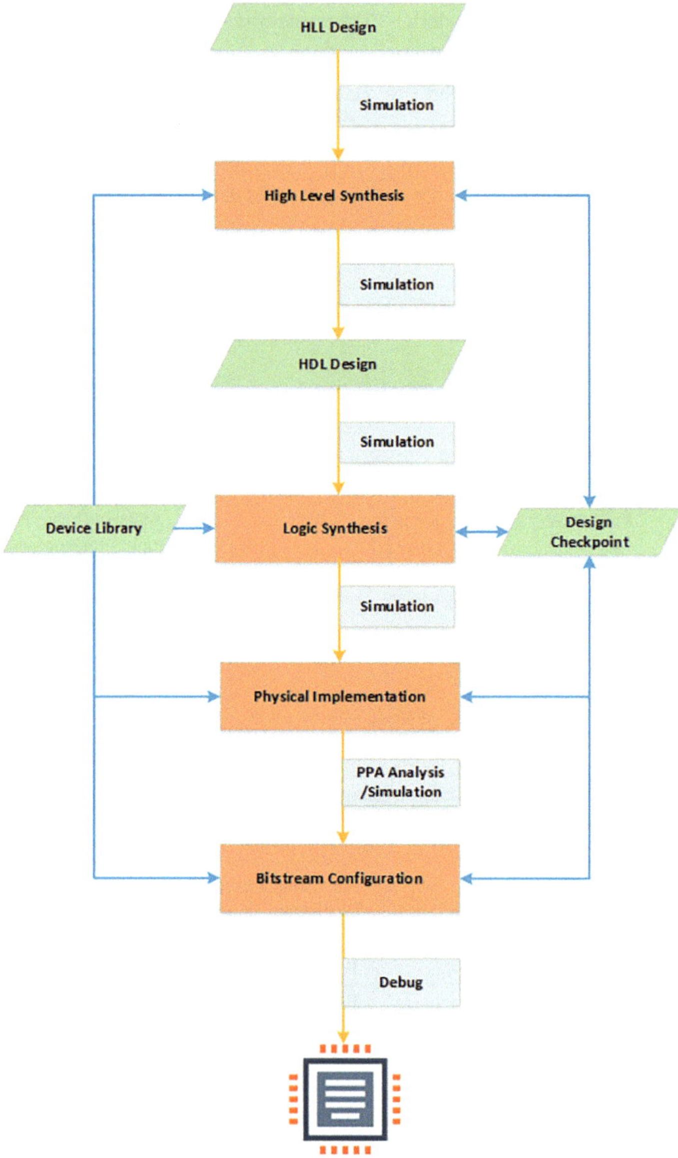

Fig. 1.14 Vanilla FPGA application design EDA flow

1.2.3 FPGA Application Design EDA

The traditional FPGA application design EDA process includes high-level synthesis, logic synthesis, physical implementation, bitstream configuration, and simulation/debugging. The FPGA device data (Device Library) and design data (Design Checkpoint) are the two major data sources for the EDA system. More specifically, high-level synthesis converts the user's high-level abstract description (High-Level Language, HLL) design into a low level hardware description (Hardware Description Language, HDL) design; logic synthesis converts the low level hardware description design into a design database that contains a design netlist composed of FPGA design units and their interconnections; physical implementation further decides how the design is physically implemented on the target FPGA; bitstream configuration converts the final implemented design into an bitstream of a specific format, and then downloads it into the target device according to the configuration protocol. After each process above is done, the simulation tools and PPA analysis tool can be invoked to sign-off the design. Simulation and debugging activities are carried out throughout the application design process to ensure that the original idea of the designer is implemented properly.

There is a lot of legacy practices constructing FPGA application design EDA flow both from industrial and academic world. Table 1.7 lists some of the most iconic frameworks (Fig. 1.14).

References

1. IEEE, 25 microchips that shook the world (2009). https://spectrum.ieee.org/25-microchips-that-shook-the-world
2. IEEE, Chip hall of fame: Xilinx xc2064 FPGA (2017). https://spectrum.ieee.org/chip-hall-of-fame-xilinx-xc2064-fpga
3. M. Abusultan S.P. Khatri, Exploring static and dynamic flash-based FPGA design topologies, in *2016 IEEE 34th International Conference on Computer Design (ICCD)* (2016), pp. 416–419
4. N. Rezzak, J.-J. Wang, D. Dsilva, N. Jat, TID and see characterization of Microsemi's 4th generation radiation tolerant rtg4 flash-based FPGA, in *2015 IEEE Radiation Effects Data Workshop (REDW)* (2015), pp. 1–6
5. P.R.C. Villa, R.C. Goerl, F. Vargas, L.B. Poehls, N.H. Medina, N. Added, V.A.P. de Aguiar, E.L.A. Macchione, F. Aguirre, M.A.G. da Silveira, E.A. Bezerra, Analysis of single-event upsets in a Microsemi proasic3e FPGA, in *2017 18th IEEE Latin American Test Symposium (LATS)* (2017), pp. 1–4
6. J. Greene, S. Kaptanoglu, W. Feng, V. Hecht, J. Landry, F. Li, A. Krouglyanskiy, M. Morosan, V. Pevzner, A 65 nm flash-based FPGA fabric optimized for low cost and power, in *Proceedings of the 19th ACM/SIGDA International Symposium on Field Programmable Gate Arrays (FPGA '11)* (ACM, New York, NY, USA, 2011), pp. 87–96
7. H.S.P. Wong, H.Y. Lee, S. Yu, Y.S. Chen, Y. Wu, P.S. Chen, B. Lee, F.T. Chen, M.J. Tsai, Metal-oxide RRAM, Proc. IEEE **100**(6), 1951–1970 (2012)
8. D. Apalkov, B. Dieny, J.M. Slaughter, Magnetoresistive random access memory. Proc. IEEE **104**(10), 1796–1830 (2016)

9. X. Tang, E. Giacomin, G.D. Micheli, P.E. Gaillardon, Circuit designs of high-performance and low-power RRAM-based multiplexers based on 4T(ransistor)1R(RAM) programming structure. IEEE Trans. Circ. Syst. I: Regul. Pap. **64**(5), 1173–1186 (2017)
10. B. Govoreanu, G.S. Kar, Y.Y. Chen, V. Paraschiv, S. Kubicek, A. Fantini, I.P. Radu, L. Goux, S. Clima, R. Degraeve, N. Jossart, O. Richard, T. Vandeweyer, K. Seo, P. Hendrickx, G. Pourtois, H. Bender, L. Altimime, D.J. Wouters, J.A. Kittl, M. Jurczak, $10 \times 10\,\text{nm}^2$ $H_f/H_f O_x$ crossbar Resistive RAM with excellent performance, reliability and low-energy operation, in *2011 International Electron Devices Meeting*, Dec 2011, pp. 31.6.1–31.6.4
11. X. Tang, G. Kim, P.-E. Gaillardon, G. De Micheli, A study on the programming structures for RRAM-based FPGA architectures. IEEE Trans. Circ. Syst. I: Regul. Pap. **63**(4), 503–516 (2016)
12. O. Turkyilmaz, S. Onkaraiah, M. Reyboz, F. Clermidy, C.A. Hraziia, J. Portal, M. Bocquet, RRAM-based FPGA for "Normally off, instantly on" applications, in *2012 IEEE/ACM International Symposium on Nanoscale Architectures (NANOARCH)*, July 2012, pp. 101–108
13. J. Cong, B. Xiao, FPGA-RPI: a novel FPGA architecture with RRAM-based programmable interconnects. IEEE Trans. Very Large Scale Integr. (VLSI) Syst. 2**2**(4), 864–877 (2014)
14. R. Rajaei, Radiation-hardened design of nonvolatile MRAM-based FPGA. IEEE Trans. Magn. **52**(10), 1–10 (2016)
15. X. Tang, P.E. Gaillardon, G.D. Micheli, A high-performance low-power near-V_t RRAM-based FPGA, in *2014 International Conference on Field-Programmable Technology (FPT)*, Dec 2014, pp. 207–214
16. S. Tanachutiwat, M. Liu, W. Wang, FPGA based on integration of CMOS and RRAM. IEEE Trans. Very Large Scale Integr. (VLSI) Syst. **19**(11), 2023–2032 (2011)
17. Y.Y. Liauw, Z. Zhang, W. Kim, A. El Gamal, S.S. Wong, Nonvolatile 3D-FPGA with monolithically stacked RRAM-based configuration memory, in *Solid-State Circuits Conference Digest of Technical Papers (ISSCC), 2012 IEEE International* (IEEE, 2012), pp. 406–408
18. Y.C. Chen, W. Wang, H. Li, W. Zhang, Non-volatile 3D stacking RRAM-based FPGA, in *22nd International Conference on Field Programmable Logic and Applications (FPL)*, Aug 2012, pp. 367–372
19. K. Huang, R. Zhao, W. He, Y. Lian, High-density and high-reliability nonvolatile field-programmable gate array with stacked 1D2R RRAM array. IEEE Trans. Very Large Scale Integr. (VLSI) Syst. **24**(1), 139–150 (2016)
20. X. Wu, 3d-IC technologies and 3d FPGA, in *2015 International 3D Systems Integration Conference (3DIC)* (2015), pp. KN1.1–KN1.4
21. AMD, Versal ACAP configurable logic block architecture manual. https://docs.xilinx.com/r/en-US/am005-versal-clb/CLB-Architecture (2021)
22. Intel, Intel stratix 10 logic array blocks and adaptive logic modules user guide (2022). https://www.intel.com/content/www/us/en/docs/programmable/683699/current/lab.html
23. A. Boutros, V. Betz, FPGA architecture: principles and progression. IEEE Circ. Syst. Magazine **21**(2), 4–29 (2021)
24. S. Yazdanshenas, V. Betz, Automatic circuit design and modelling for heterogeneous FPGAs, in *2017 International Conference on Field Programmable Technology (ICFPT)* (2017), pp. 9–16

Part II
FPGA Data Modeling

Chapter 2
Device (Chip Design) Modeling

Abstract This chapter provides the principles and implementations of FPGA device (chip design) modeling. FPGA device information can be derived from the output of chip design data and then will become the input for application design. There are two practices in the FPGA EDA full flow that can share a common set of device models, one is for architecture exploration in the chip design stage, the other is for implementing end user's circuit in the application design stage.

2.1 Device Description Levels

2.1.1 Abstract Levels

FPGA chip design is the output of the chip design EDA stage, it begins with a design description, which is quite similar to ASIC chip design. FPGA chip design can be described at different abstraction levels: natural level [1], high level [2], low level [3], machine level [4], and physical level, forming a pyramid (Fig. 2.1) from top to bottom.

FPGA device library comes from the chip design to some extent, yet is another different concept. It carries the FPGA chip design data that only will be used in the application design EDA stage. Therefore, describe an FPGA device at natural level, high level, or physical level is not yet wide spread in industry due to the immaturity of related tools, most of the existing research works about FPGA device description mainly stay at low level and machine level (Fig. 2.2).

1. Physical-Level Description
   The device physical-level description refers to the layout of targeted technology, representing geometric shapes, text labels, etc. It is the final interface between chip designer and foundry. GDSII is the de facto industry standard format to carry messages at this level.
2. Machine-Level Description
   The device machine-level description refers to the structure of underlying bit-stream. It can be further divided into two sub-level descriptions: the "logical" description expresses each configuration bit's logical address—the correlation

K. Tu et al., *FPGA EDA*, https://doi.org/10.1007/978-981-99-7755-0_2

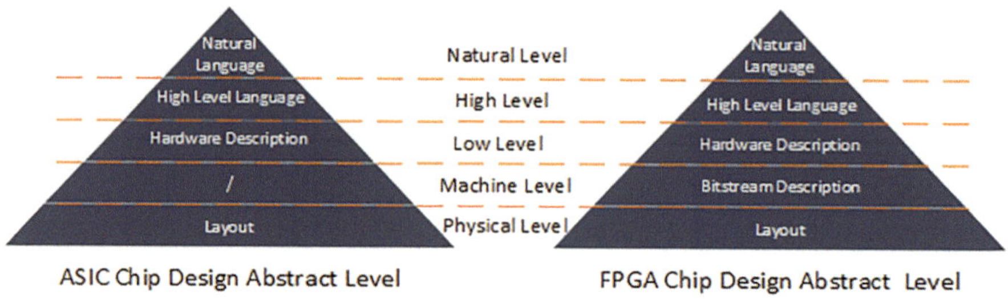

Fig. 2.1 Abstract levels of chip design

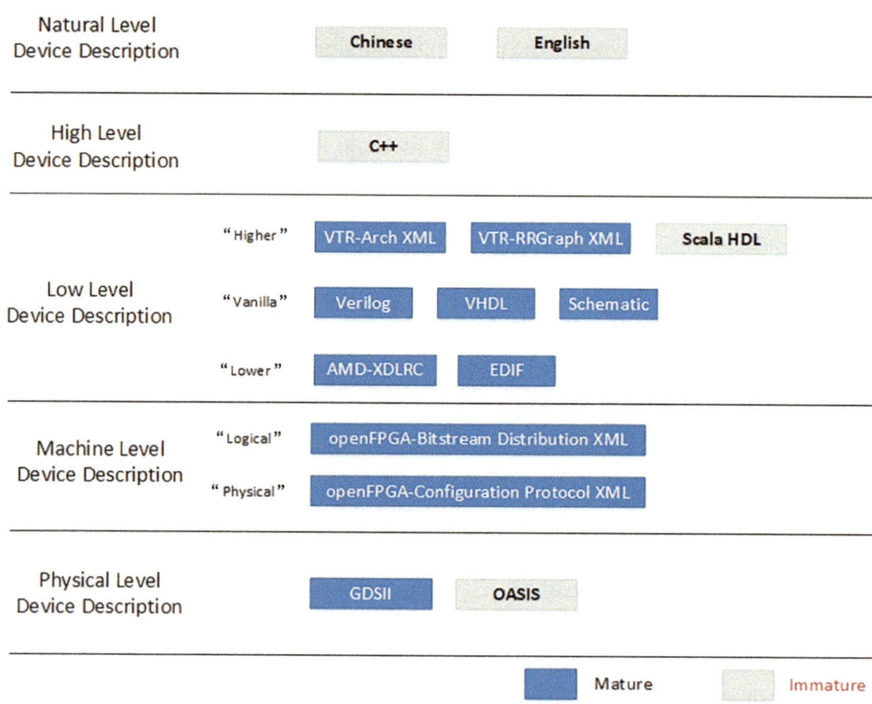

Fig. 2.2 FPGA device description abstract levels

between bits and hardware units; the "physical" description presents each configuration bit's physical address—the bit physical organization format decided by configuration protocol.

3. Low-Level Description

   The device low-level description refers to the characterization of hardware, and three sub-level descriptions are demarcated: the "vanilla" description, the "higher" description and the "lower" description.

   At "vanilla" level, traditional vendor-neutral hardware descriptions (such as VHDL, Verilog, Schematic, or SPICE) are very common to represent the hard-

Table 2.1 Feature comparison of XML and JASON

Feature	XML	JASON
Format	Format that has tags to define elements	Format written in JavaScript
Data storage	As a tree structure	Like a map with key value pairs
Speed	Bulky and slow in parsing, leading to slower data transmission	Very fast as the size of file is considerably small, faster parsing by the JavaScript engine and hence faster transfer of data
File size	Size is bulky, the tag structure makes it huge and complex to read	Compact and easy to read, no redundant or empty tags or data, making the file look simple
Security	Document Type Definition (DTD) validation and external entity expansion are enabled by default, making structures disposed to some attacks	Safe at all times, although it might be more at risk when JSONP (JSON with Padding) is used since it can result in a Cross-Site Request Forgery (CSRF) attack

ware details.

At "higher" level, more abstracted hardware descriptions (such as XML, JASON) (Table 2.1) are very suitable for FPGA structure exploration in the early stage of design. Because if the promising structure is not ideal in the subsequent verification, designers can modify it at a higher level to quickly establish a new fine-tuned structure. However, description from this level may not fully reflect all special details of the hardware.

At "lower" level, more detailed hardware descriptions (such as AMD-XDLRC, EDIF) is used to specify detailed units in the FPGA device.

4. High-Level Description
   The device high-level description refers to algorithmic languages (such as C++). High-level synthesis (HLS) EDA engine is needed to convert descriptions at this level into lower level.
5. Natural-Level Description
   In addition to the above levels of device descriptions, industry's pursuit of higher abstraction (to a certain extent, also to reduce the labor cost of circuit design) has never stopped. The final form of device description is the language used by humans (English, Chinese, etc.). Natural language is the main tool for human communication and thinking, it is the crystallization of human wisdom. Natural language processing is also one of the most difficult problems in artificial intelligence [5].

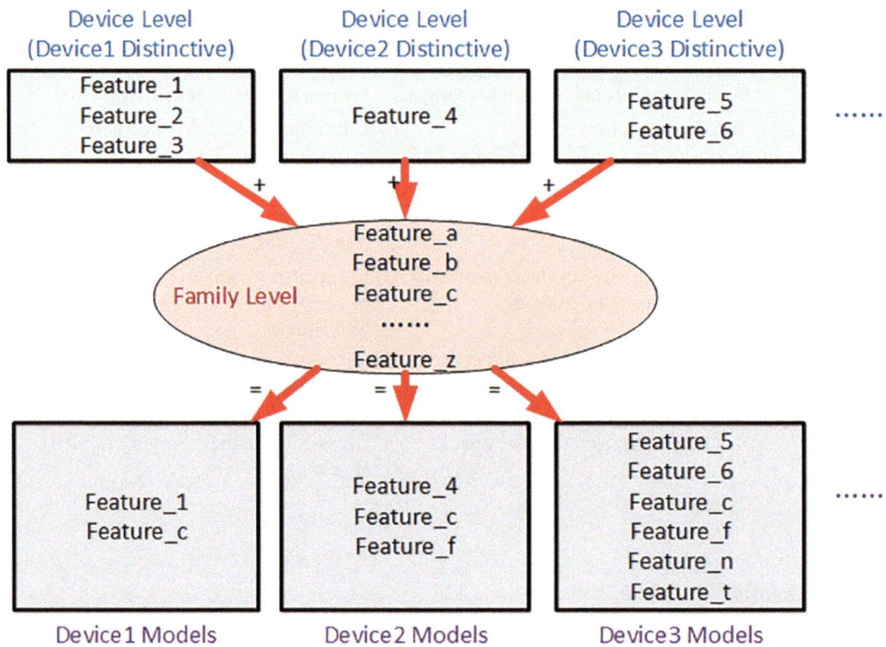

Fig. 2.3 FPGA device description reuse levels

2.1.2 *Reuse Levels*

Considering the design reusability, modern FPGA vendors usually design a series of chips based on the same technology platform (process node, infrastructure features, cell libraries, etc.), which we call it "family". Under the same family, each device has a unique resource scale. Take AMD's FPGA as an example, there are dozens of devices in its Virtex-7 family that shares the same technology platform.

Considering there are common components (IPs) in an FPGA device from the same family, in order to simplify the complexity and improve the efficiency, the design features can be accordingly modeled at two levels: family level that shared by the entire family, and device level that only suitable for a specific single device (Fig. 2.3).

2.2 Device Model Classifications

The device models of an FPGA partly come from the chip design. The "device library", built from these models, contains the device architecture-related EDA information that serves the application design flow. Device models can be categorized into

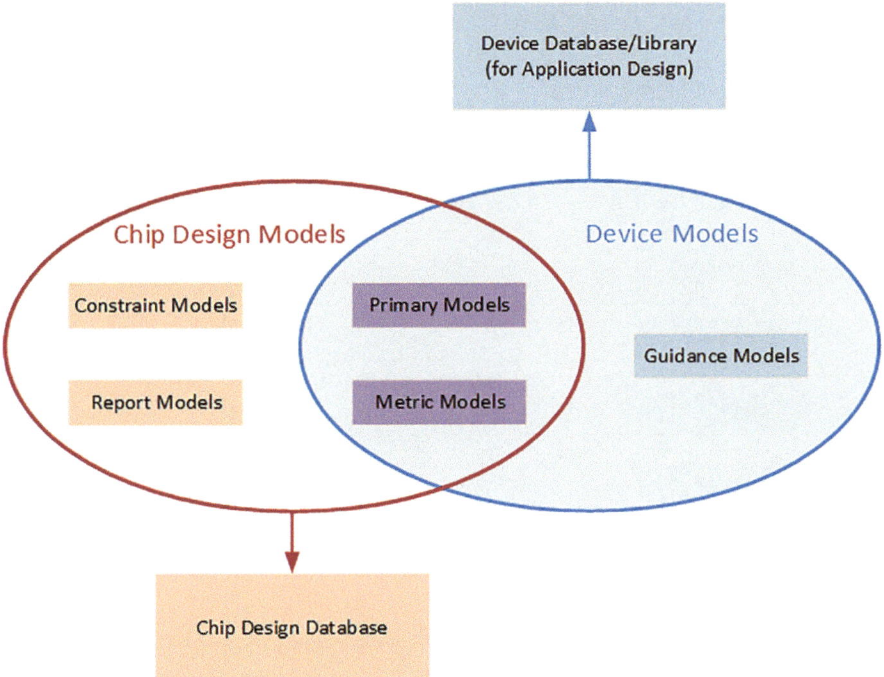

Fig. 2.4 FPGA device model classifications

several classes: primary, metric, and guidance (Fig. 2.4). When doing chip design, constraint models and report models are required, however, they are irrelevant with the device modeling.

2.2.1 Primary Class

The device primary models include logical resource structure model and configuration bit structure model. These models are essential inputs to application design EDA, and can be derived from the chip design process.

1. Logic Resource Structure (LRS) Model
   The logic resource structure of an FPGA device contains the core, the package, and the interconnect among them (Fig. 2.5).
   The FPGA core has a traditional structural hierarchy: core-tile-site-primitive-gate-transistor. In chip design EDA, the full-custom method designs from the bottom transistor-level, while the semi-custom method designs from the relatively upper gate-level. In application design EDA, the situation is similar to semi-custom chip design, gate-level is the lowest noteworthy level since that is where logic synthesis

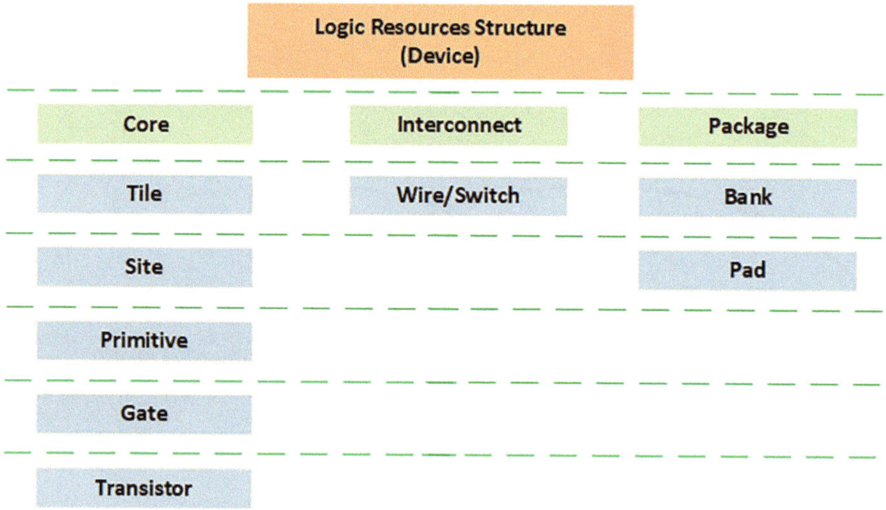

Fig. 2.5 FPGA device logical resource structure hierarchy

could possibly reach. It is worth mentioning that the design of modern clock networks could partition *tile* resources into different "regions", and some multi-chip packaged devices also define each stacked silicon as a "die", nonetheless, both these "dies" and "regions" are made of *tiles*. In order to simplify the illustration, we uses *tile* as the first-level sub-unit of a device.

The FPGA package has a much simpler structural hierarchy: package-bank-pad. A bank is a group of I/O pads that share a common resource such as one power supply or one output current reference, each group can independently support different I/O standards that can adapt to different electrical characteristics. At pad-level, pad arrangement/layout is required to make the package information intact.

The FPGA interconnect resources organically combine all kinds of logic units together. These programmable interconnect resources can be abstracted as a network composed of wires and programmable switches. Wire is the carrier for signal transmission, and switch controls the flow direction of signals by switching on and off. They together form the device's routing architecture.

2. Configuration Bit Structure (CBS) Model

The configuration bit structure of an FPGA device can be defined from two perspectives: logical and physical.

Logical CBS gives every bit a "logical address", that is, which logic resource it belongs to (Fig. 2.6).

Physical CBS gives every bit a "physical address", that is, which position it lies in the final bitstream sequence according to the configuration protocol.

Configuration protocol depends on the circuitry designed to program an FPGA. It could be in different structures based on the application context, providing differ-

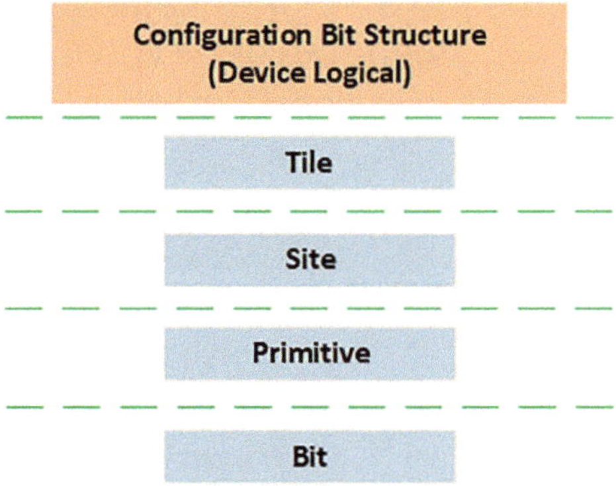

Fig. 2.6 FPGA device configuration bit structure hierarchy (logical)

ent trade-offs between speed and area. In industry, FPGA devices from the same family usually share the same configuration protocol. Here are some representative structures:

a. Chain-based
   In this structure, configurable memories are connected in one or multiple chains and bitstream is loaded serially to program the FPGA.
b. Unit-based
   The configuration memory is organized by logical units (such as *tiles* and *sites*) in which it resides, each configuration memory can be accessed by an address decoder. Due to the hierarchical structure of logical units, unit-based physical bit structure is also hierarchical.
c. Region-based
   The configuration memory is organized in the form of a matrix area on the FPGA chip, and each configuration memory can be accessed through the Bit Line/Word Line address decoder.

Bitstream is more than the bits to configure an FPGA, it also contains certain human-readable fields (meta) to describe those bits and an assembly-like instruction set (command) to guide the FPGA configuration process. Packet-Frame-Bit hierarchy, borrowed from networking OSI(Open Systems Interconnection) model (Fig. 2.7), is used for the configuration data:

a. Packet
   Packet is the basic unit of communication between a source and a destination in a network. In the OSI model, packets are data units within the network layer.
b. Frame
   Just like packets, frames are small parts of a message in the network. The main difference between a packet and a frame is the association with the OSI layers—frames are data units within the data link layer.

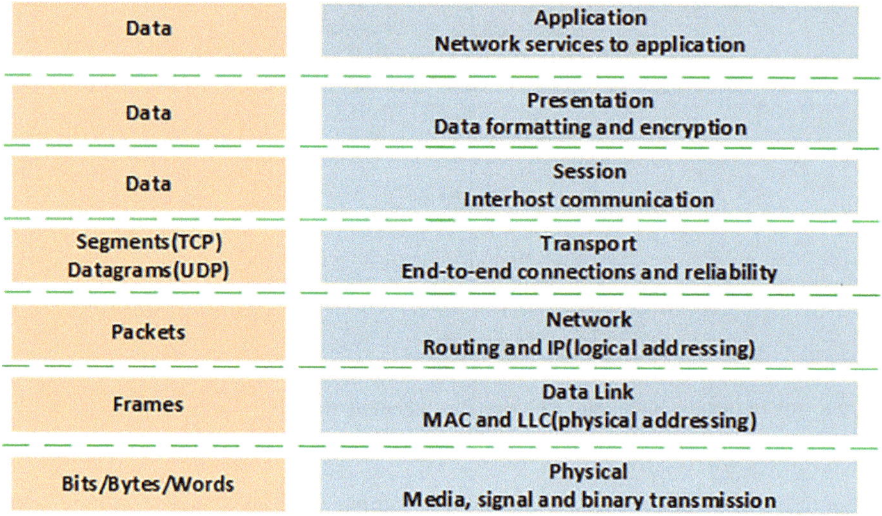

Fig. 2.7 Communication data in OSI model

c. Bit/Byte/Word
   Configuration memory stores the data in bits at the bottom level. Each bit stores the value either 0 or 1. Each byte has 8 bits, and each word can has a different length in different systems (such as 8/16/32 bits). Registers are used to store a small piece of information (byte/word) while doing the calculations or processing the data. It is helpful to improve the performance of the system while doing the calculation or processing.

As discussed above, FPGA physical bit structure has a general hierarchy in (Fig. 2.8).

After the logical and physical structure are proper defined, the logical–physical address correlation for each bit is then established (Fig. 2.9).

2.2.2 *Metric Class*

There are several metrics that must be concerned during the FPGA design process, and PPA (Power/Performance (Timing)/Area) is the most critical ones among them. In order to monitor these metrics accurately and sign off effectively, analysis is a conventional activity that every EDA system cannot get around. There are two ways to do the analysis: measurement and estimation.

Measurement can only be carried out after the FPGA chip is configured and ready to work. It is essentially a testing procedure that could achieve the most accurate analysis results, at the price of extra efforts such as time-consuming full chip design, manufacturing and external test instrumentation setups.

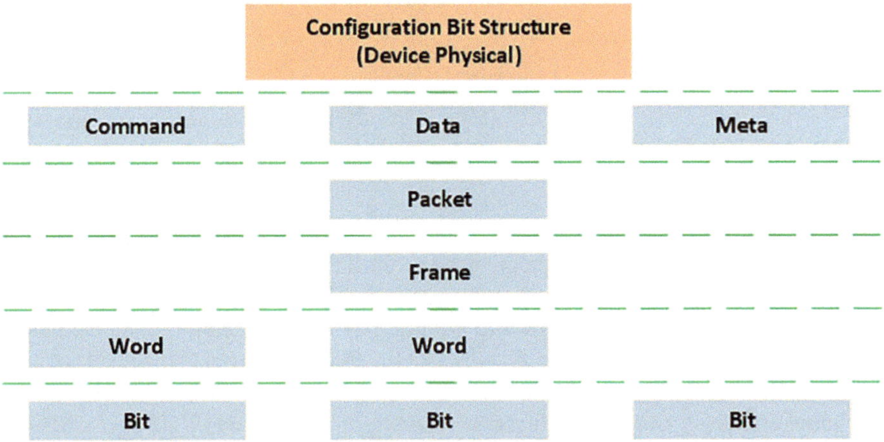

Fig. 2.8 FPGA device configuration bit structure level (physical)

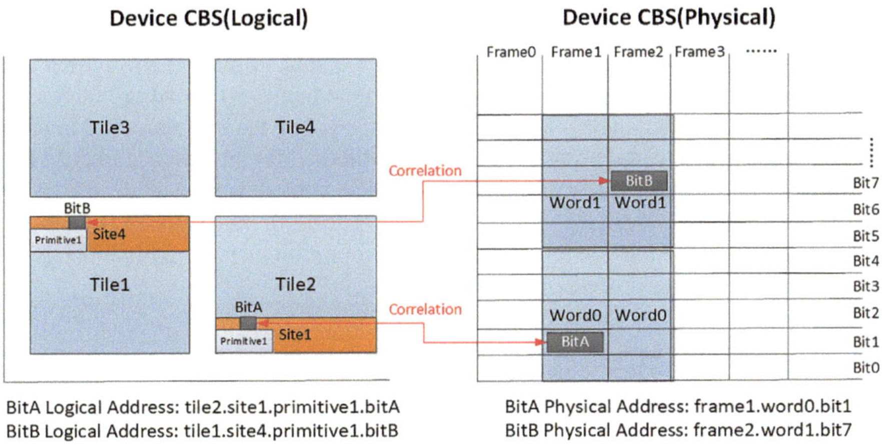

Fig. 2.9 FPGA device configuration bit correlation

Estimation, on the other hand, can be performed before the chip design is complete or even at the very early design stage. Models are used to approximate the result, which makes estimation a less costly and fairly efficient alternative to the measurement solution.

For FPGA, there are estimations and measurements at different stages to ensure the design objectives are met (Fig. 2.10). In terms of EDA part, we only discuss the PPA models used by the estimation tasks here.

Metric models are software-based representation of the physical parameters in the FPGA. There are different statuses for metric models in the FPGA design cycle: advance, preliminary, and final. Advance models are typically available soon after the device design specifications are frozen and may change as silicon characterization

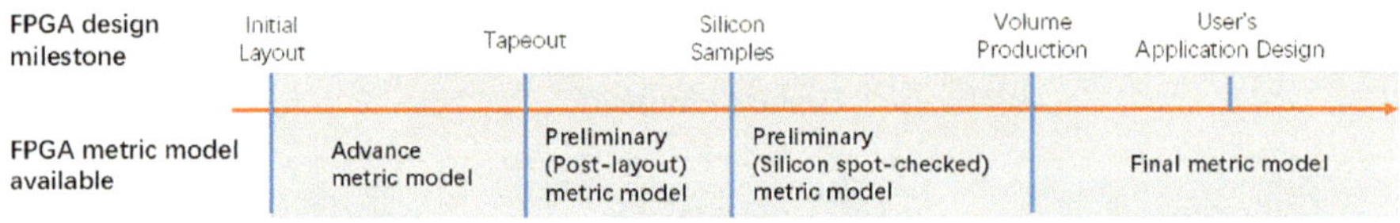

Fig. 2.10 FPGA metric models during the development life cycle

data becomes available. Preliminary models are based on early production silicon and all the units in the device are characterized. Final models correlate to production devices with thousands of designs and are not expected to change. Besides, metric models can be constructed at different abstract levels (such as transistor-level, switch-level, gate-level, register transfer-level, high-level, etc.). Higher abstractions allow for quicker estimations, but with reduced accuracy. For large-scale FPGAs, estimations only at gate-level or lower could guarantee the sufficient accuracy, however, running at bottom transistor-level will also make calculation time intolerable. The generation of FPGA metric model is also a complicated process (Fig. 2.11) is a typical flow from Intel.

Except for how these metric models are represented in the FPGA architecture file, another important issue, which will be detailed in the following chapters—[Part. III], is how these models are properly built. Among the best practices to build these metric models, simulation/layout-based methods offer the best accuracy, while equation-based methods trade it off with efficiency (thus mostly used during architecture exploration). The main problem with power/timing model for FPGAs is that the power/timing depends on inputs and configuration bits which maintain circuit's behavior, temperature, process, voltage, etc.

1. Power Model
   Power model is prerequisite for power analysis and power-aware EDA engines [7].

   a. Equation-based
      In equation-based methods, power model is build by analytical calculations via predefined equations [8, 9].
   b. Simulation-based
      In simulation-based methods, as its name suggests, power model is obtained by simulations. The information in the model (including current and voltage values, capacitance, etc.) is different depending on the abstraction level of the simulator. SPICE is the most used transistor-level simulator [10].

2. Timing Model
   Timing model is essential for timing analysis and timing-driven EDA engines.

   a. Equation-based
      In equation-based methods, the Elmore delay is the most frequently employed during architecture exploration[11–15].

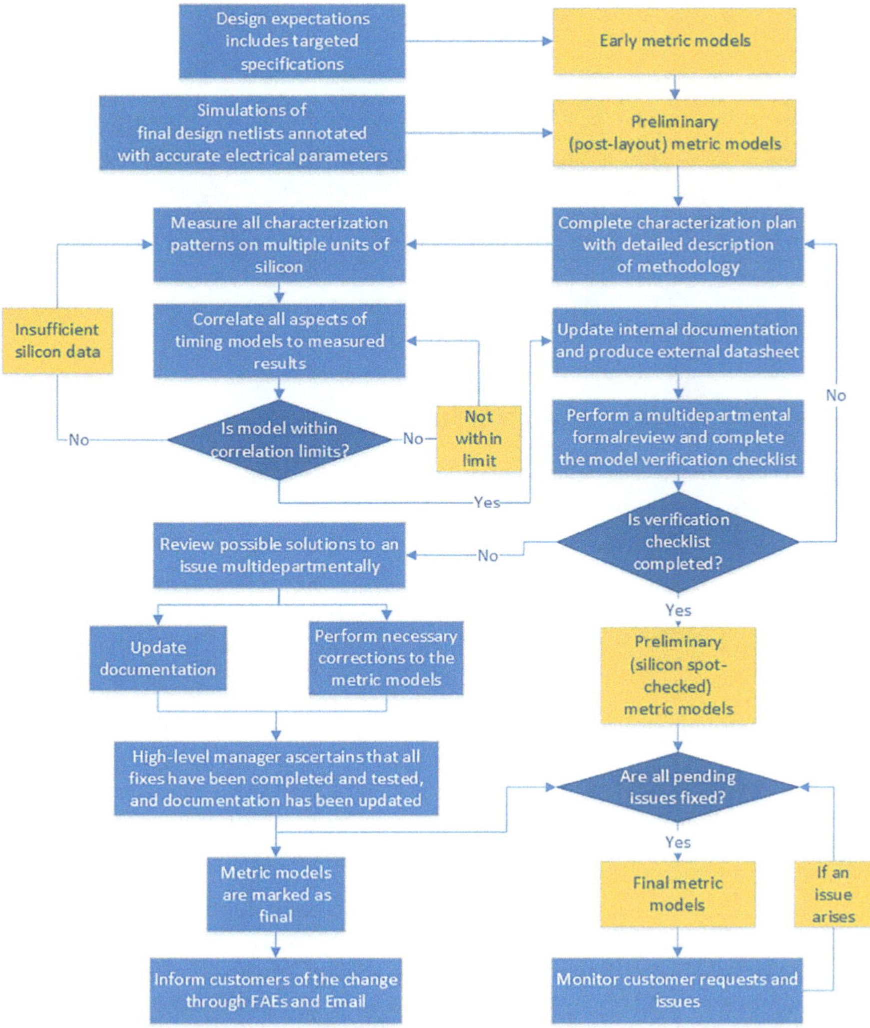

Fig. 2.11 Typical flow of FPGA metric models generation [6]

b. Simulation-based

In industry, circuit simulator is used to get the final timing metrics. The circuit simulator (such as SPICE) extracts all electrical data, such as capacitance and resistance, and all nonlinear and linear components, to determine the expected delays.

3. Area Model

Just like ASIC design, chip area is a metric that must be concerned. Advanced process nodes can shrink the area dramatically, more specific, parameters such

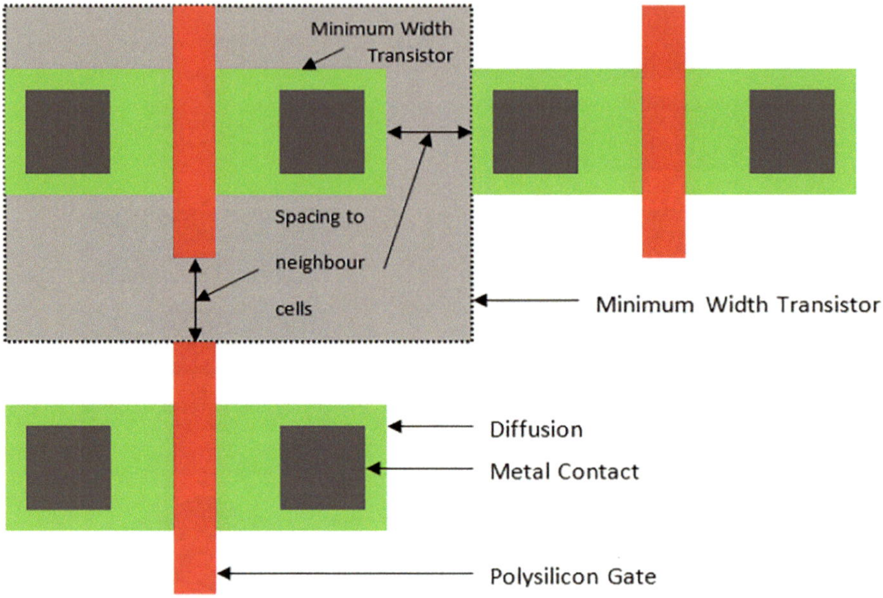

Fig. 2.12 Minimum-width transistor area model [16]

as the transistor size and the number of metal layers are taken into account to estimate the chip area [16]. This estimation only occurs at chip design stage; once the FPGA is manufactured, the end user cannot change the chip area; however, they can still evaluate how many logic resources their design will utilize.

a. Equation-based
   In equation-based methods, minimum-width transistor area model (MWTA) is popular for academia. A minimum-width transistor is defined as the smallest possible contactable transistor for a specific process technology and one minimum-width transistor area is the area of this transistor plus the spacing to neighboring transistors (Fig. 2.12).
   For example, a 1× (unit-sized) CMOS inverter consists of two minimum width transistors (a PMOS pull-up, and NMOS pull-down). This model ignores the differences between different semiconductor processes, which could lead to a rough result [17].
b. Layout-based
   In layout-based methods, FPGA chip area can be metered precisely by layout EDA tools (such as Cadence Virtuoso).

2.2.3 Guidance Class

Some application design in EDA engine has their dedicate device architecture dependencies that cannot directly obtained from the chip design process. These special device characteristics fall into the last class: guidance. Models of guidance class only serve the application design process.

1. Packing/Placement Guidance Model
   Except for device LRS, packing/placement guidance information is required for the packing/placement engine. This information (such as which design units can be packed together, placed together/at specific location, etc.) should be modeled in advance.
   Modern FPGA device units (*tiles/sites*, etc.) are designed to operate in various modes, so as to provide best performance for different applications. Packing/placement engines will work out optimized solutions for each design unit to accommodate in the proper device unit by choosing these modes wisely.

2. Routing Guidance Model
   Except for device LRS, routing guidance information is required for the routing engine. This information (such as logically equivalent of routing nodes and edges, etc.) should be modeled in advance.

3. Bitstream Generation Guidance Model
   The bitstream generation guidance model is the device architecture information that a bitstream generator must concern. Programmable point (PP) is the most important concept in this model. PP, usually controlled by one or several configuration bits, is a programming unit in FPGA that could run a basic function independently. Each PP can be programmed into several modes, and each possible programming state is called a bitstream configuration mode of the PP.
   In terms of resource type they belong, PPs can be divided into two categories:

   a. Programmable Logic Point (PLP)
      PLP represents the programming point within a logical unit resource, and its bitstream configuration mode can be defined as a data pair: parameter and its multi-option values.
   b. Programmable Interconnect Point (PIP)
      PIP represents the programming point of interconnect resources, and its bitstream configuration mode can also be defined as a data pair: output pin/wire and its multi-option input pin/wire.

   Configuration modes for a PP can be described in two ways:

   a. Enumeration
      When the configuration modes of a programming point are relatively few, and the configuration value under each mode configuration is a certain value, its configuration modes then can be listed by enumerating all the possibilities.

b. Formulation
   When the configuration modes of a programming point are too many to be enumerated (such as the mask value of LUT, the duty cycle of PLL, each possible value can be regarded as a configuration mode). Under this circumstances, only one mode is listed, and the configuration value can be calculated by a pre-defined formula.

2.3 Device Model Implementations

The previous section listed all the device model classes: primary class, metric class and guidance class. In this section, we will present typical implementation practices for each model at proper level (Table 2.2).

2.3.1 Logic Resource Structure Model

FPGA device LRS model includes device, core, package, and interconnect models.

1. Device

   1. Device basic information
      Device basic information includes: device name, family name, core name, package name, speed grade, temperature grade, etc.
      Implementation Example: (Fig. 2.13)

2. Core

   a. Core information
      Core information contains: core name, family name, the height/width of the array, and the coordination of its internal tile-level resources. Tile/Site information can be similar.
      Implementation Example: (Fig. 2.14)

Table 2.2 Comparison of FPGA device model implementations

Model name	Abstract level	Reuse level	Class
Logic resource structure	Low	Hybrid	Primary
Configuration bit structure	Machine	Hybrid	Primary
Power	Low	Family	Metric
Timing	Low	Family	Metric
Area	Low	Family	Metric
Packing/placement guidance	Low	Family	Guidance
Routing guidance	Low	Family	Guidance
Bitstream generation guidance	Low	Family	Guidance

```
<!-- Device basic information modelled in XML -->
<device device_name="string" family_name="string"
core_name="string" package_name="string" speed_grade="string"
temperature_grade="string"/>
```

Fig. 2.13 Example of XML syntax for device basic information

```
<!-- Core information modelled in XML -->
<core core_name="string" family_name="string" height="int"
width="int">
  <detail x="int" y="int" tile_name="string" tile_type="string"/>
  <detail x="int" y="int" tile_name="string" tile_type="string"/>
  ......
</core>
```

Fig. 2.14 Example of XML syntax for core information

```
<!-- Tile information modelled in XML -->
<tile tile_name="string" height="int" width="int">
  <detail x="int" site_name="string" site_type="string"/>
  <detail x="int" site_name="string" site_type="string"/>
  ......
</tile>
```

Fig. 2.15 Example of XML syntax for tile information

```
<!-- Site information modelled in XML -->
<site site_name="string">
  <detail x="int" primitive_name="string" primitive_type="string"/>
  <detail x="int" primitive_name="string" primitive_type="string"/>
  ......
</site>
```

Fig. 2.16 Example of XML syntax for site information

b. Tile information
   Tile is the sub-module of core.
   Implementation Example: (Fig. 2.15)
c. Site information
   Site is the sub-module of tile.
   Implementation Example: (Fig. 2.16)

3. Package

   a. Package information
      Package information contains: package name, family name, the height/width of the array, and the coordination and bank name of its pad resources.
      Implementation Example: (Fig. 2.17)

```
<!-- Package information modelled in XML -->
<package package_name="string" family_name="string" height="int"
width="int">
  <detail x="int" y="int" pad_name="string" pad_type="string"
bank_name="string"/>
  <detail x="int" y="int" pad_name="string" pad_type="string"
bank_name="string"/>
  ......
</package>
```

Fig. 2.17 Example of XML syntax for package information

```
<!-- Wire/Switch information modelled in XML -->
<interconnect>
  <wire input="lut_4.out" output="ff.D"/>
  <wire input="ble.in" output="lut_4.in"/>
  <switch type="mux" input="ff.Q lut_4.out" output="ble.out"/>
  <wire input="ble.clk" output="ff.clk"/
  ......
</interconnect>
```

Fig. 2.18 Example of XML syntax for interconnect information

4. Interconnect

   a. Wire/Switch information
      Input and output information are essential for interconnects.
      Implementation Example: (Fig. 2.18)

2.3.2 *Configuration Bit Structure Model*

FPGA device CBS model includes logical information, physical information, and the correlation between them.

1. Logical CBS information
   Logical CBS information contains: each memory port's logical address.
   Implementation Example (for a tile): (Fig. 2.19)
2. Physical CBS information
   Physical CBS information contains: each memory port's physical address.
   Implementation Example: (Fig. 2.20)
3. CBS correlation information
   CBS correlation information contains: each memory port's logical–physical address correlation.
   Implementation Example (for a tile): (Fig. 2.21)

```
<!-- Logical CBS information modelled in XML -->
<tile_cbs_logical type="string" bit_num="int">
  <site1_cbs_logical type="string" bit_num="int">
    <primitive1_cbs_logical type="string" bit_num="int">
      <port="string">
      <port="string">
      ......
    </primitive1>
    ......
  </site1>
  ......
</tile_cbs_logical>
```

Fig. 2.19 Example of XML syntax for logical CBS information of a tile

```
<!-- Physical CBS information modelled in XML -->
<device_cbs_physical family="string" name="string" bit_num="int">
  <meta name="string" bit_num="int">
  <packet name="string" bit_num="int">
    <frame name="string" bit_num="int">
      <bit name="string"/>
      <bit name="string"/>
      ......
    </frame>
    ......
  </packet>
  ......
</device_cbs_physical>
```

Fig. 2.20 Example of XML syntax for physical CBS information of a device

```
<!-- CBS correlation modelled in XML -->
<tile_correlation type="string">
  <correlation logical_address="string" physical_address="string"/>
  <correlation logical_address="string" physical_address="string"/>
......
</tile_correlation>
```

Fig. 2.21 Example of XML syntax for CBS correlation information of a tile

2.3.3 Power Model

Power model in the architecture description refers to the reference power characteristics of each FPGA units.

Implementation example: VTR (Fig. 2.22)

The most basic representation specifies both the dynamic and static power of an FPGA unit as absolute values (in Watts). This is done using the following construct:

Implementation example: Industrial

```
<!-- Power characteristics modelled in XML -->
<power method="absolute">
  <primitive type="string">
    <dynamic_power power_per_instance="1.0e-16"/>
    <static_power power_per_instance="1.0e-16"/>
  </primitive>
  ......
</power>
```

Fig. 2.22 Example of XML syntax for power information of a primitive

```
<!-- Setup time modelled in XML -->
<T_setup value="float" port="string" clock="string"/>
```

Fig. 2.23 Example of XML syntax for setup time model

```
<!-- Hold time modelled in XML -->
<T_hold value="float" port="string" clock="string"/>
```

Fig. 2.24 Example of XML syntax for hold time model

1. Liberty (.lib) [18]
   The Liberty (LIB) format from Synopsys is an ASCII file that describes an FPGA unit's characterized data in a standard way. The Liberty model contains power data such as leakage power, internal power, etc.

2.3.4 Performance (Timing) Model

Timing model in the architecture description refers to the reference delay characteristics of each FPGA units.

Implementation example: VTR

1. Setup timing model (Fig. 2.23)
   Attributes:
   value—The setup time value.
   port—The port name the setup constraint applies to.
   clock—The port name of the clock the setup constraint is specified relative to.
2. Hold timing model (Fig. 2.24)
   Attributes:
   value—The hold time value.
   port—The port name the setup constraint applies to.
   clock—The port name of the clock the setup constraint is specified relative to.
3. Clock to Q timing model (Fig. 2.26)
   Attributes:

```
<!-- Clock to Q time modelled in XML -->
<T_clock_to_Q max="float" min="float" port="string"
clock="string"/>
```

Fig. 2.25 Example of XML syntax for constant timing model

```
<!-- Constant delay modelled in XML -->
<delay_constant max="float" min="float" in_port="string"
out_port="string"/>
```

Fig. 2.26 Example of XML syntax for clock to Q time model

max—The maximum clock-to-Q delay value.
min—The minimum clock-to-Q delay value.
port—The port name the delay value applies to.
clock—The port name of the clock the clock-to-Q delay is specified relative to.

4. Constant timing model (Fig. 2.25)
   Specifies a maximum and/or minimum delay from input port to output port. Note that the path from input port to output port can be combinational or sequential.
   Attributes:
   max—The maximum delay value.
   min—The minimum delay value.
   in_port—The input port name.
   out_port—The output port name.

Implementation example: Industrial

1. Liberty (.lib) [18]
   The Liberty (LIB) format from Synopsys is an ASCII file that describes an FPGA unit's characterized data in a standard way. The Liberty model contains timing data such as setup time, hold time, recovery time, removal time, etc.
2. Parasitics (.spef) [19]
   Standard Parasitic Exchange Format (SPEF) is an IEEE standard for specifying chip parasitics. Specifically, it defines the design parasitics of a set of nets as a resistive-capacitive (RC) network, which will be used to calculate routing delay.
3. Standard Delay Format (.sdf) [20]
   Standard Delay Format (SDF) is an IEEE specification to represent circuit delays. The LIB only has the cell delays in a table form, and the SPEF file has the interconnect parasitics. SDF file combines these information and gives out accurate delays for each component in the layout database, for the given constraints.

```
<!-- Constant area modelled in XML -->
<area grid_logic_tile_area="float"/>
```

Fig. 2.27 Example of XML syntax for constant area model

```
<!-- Packing/placement guidance information modelled in XML -->
<!-- Multi-mode fracturable site definition begin -->
<site type="fsite">
<!-- Dual 5-input LUT mode definition begin -->
  <mode name="n2_lut5">
    <!-- Detailed definition of the dual 5-input LUT mode -->
  </mode>
<!-- Dual 5-input LUT mode definition end -->
<!-- 6-input LUT mode definition begin -->
  <mode name="n1_lut6">
    <!-- Detailed definition of the 6-input LUT mode -->
  </mode>
<!-- 6-input LUT mode definition end -->
</site>
```

Fig. 2.28 Example of XML syntax for packing/placement guidance model

2.3.5 Area Model

Area model in the architecture description refers to the reference area characteristics of each FPGA units.

Implementation example: VTR

The default area used by each 1×1 grid tile (in MWTAs) can be specified by (Fig. 2.27), excluding routing. It can be used for an area estimate of the amount of area taken by all the functional units.

2.3.6 Packing/Placement Guidance Model

Packing/placement guidance information contains: operation mode of all FPGA device units (deciding how many kinds of design units can be placed in this device unit).

Implementation Example: VTR (for a generic logic site) (Fig. 2.28)

A Generic Logic Site (GLS) often contains numbers of LUTs and FFs. The LUT can be fracturable, so it can be operate as either a big LUT or two smaller LUTs with shared inputs.

```
<!-- Routing guidance information modelled in XML -->
<input name="I" num_pins="33" equivalent="full"/>
<output name="O" num_pins="10" equivalent="instance"/>
```

Fig. 2.29 Example of XML syntax for routing guidance model

```
<!-- Enumeration type bitstream generation guidance information
modelled in XML -->
<primitive type="string">
  <programmable_point name="string">
    <configure_mode name="string">
      <bit port_name="string" value="int">
      <bit port_name="string" value="int">
      ......
    </configure_mode>
  </programmable_point>
</primitive>
```

Fig. 2.30 Example of XML syntax for enumeration type bitstream generation guidance model of a 4-input MUX

2.3.7 Routing Guidance Model

Routing guidance information contains: logically equivalent of RRG nodes and edges. For example, an AND gate has logically equivalent inputs because you can swap the order of the inputs and it's still correct; an adder, on the other hand, is not logically equivalent because if you swap the MSB with the LSB, the results are completely wrong. LUTs are also considered logically equivalent since the logic function (LUT mask) can be rotated to account for pin swapping.

Implementation Example: VTR (for a generic logic tile) (Fig. 2.29)

2.3.8 Bitstream Generation Guidance Model

Bitstream generation guidance information contains: configuration mode of all programmable points.

1. Enumeration
   Implementation Example: (Fig. 2.30)
2. Formulation
   Implementation Example: (Fig. 2.31)

```
<!-- Formulation type bitstream generation guidance information
modelled in XML -->
<primitive type="string">
  <programmable_point name="string">
    <configure_mode name="string">
      <bit port_name="string" value="lut_mask/16">
      <bit port_name="string" value="lut_mask%16/8">
      ......
    </configure_mode>
  </programmable_point>
</primitive>
```

Fig. 2.31 Example of XML syntax for formulation type bitstream generation guidance model of a 4-input LUT

References

1. Wikipedia, Natural language (2022). https://en.wikipedia.org/wiki/Natural_language
2. Wikipedia, High level language (2022). https://en.wikipedia.org/wiki/High-level_programming_language
3. Wikipedia, Assembly language (2022). https://en.wikipedia.org/wiki/Assembly_language
4. Wikipedia, Machine code (2022). https://en.wikipedia.org/wiki/Machine_code
5. C.C. Aggarwal, *Machine Learning for Text* (The Name of the Publisher, 2018)
6. Intel, Guaranteeing silicon performance with FPGA timing models. https://cdrdv2-public.intel.com/650314/wp-01139-timing-model.pdf
7. Y. Nasser, J. Lorandel, J.-C. Prévotet, M. Hélard, Rtl to transistor level power modeling and estimation techniques for FPGA and ASIC: a survey. IEEE Trans. Comput.-Aided Des. Integr. Circ. Syst. **40**(3), 479–493 (2021)
8. K.K.W. Poon, S.J.E. Wilton, A. Yan, A detailed power model for field-programmable gate arrays. ACM Trans. Des. Autom. Electron. Syst. **10**(2), 279–302 (2005). Available https://doi.org/10.1145/1059876.1059881
9. F. Li, Y. Lin, L. He, D. Chen, J. Cong, Power modeling and characteristics of field programmable gate arrays. IEEE Trans. Comput.-Aided Des. Integr. Circ. Syst. **24**(11), 1712–1724 (2005)
10. X. Tang, E. Giacomin, G.D. Micheli, P.-E. Gaillardon, FPGA-spice: a simulation-based architecture evaluation framework for FPGAs. IEEE Trans. Very Large Scale Integr. (VLSI) Syst. **27**(3), 637–650 (2019)
11. E. Hung, S.J.E. Wilton, H. Yu, T.C.P. Chau, P.H.W. Leong, A detailed delay path model for FPGAs, in *2009 International Conference on Field-Programmable Technology* (2009), pp. 96–103
12. Q. Liu, H. Qian, Fast and accurate circuit delay model for FPGA architectural exploration. IET Comput. Digital Tech. **11**, 12 (2016)
13. J. Lu, N. Xu, J. Yu, T. Weng, Research on cell timing modeling based on FPGA cell configurations, in *2018 2nd IEEE Advanced Information Management, Communicates, Electronic and Automation Control Conference (IMCEC)* (2018) pp. 2408–2413
14. Z.-J. Qi, Q. Duan, L.-R. Hu, X.-X. Tao, J. Wang, M. Yang, J.-M. Lai, Timing model for GRM FPGA based routing, in *2018 14th IEEE International Conference on Solid-State and Integrated Circuit Technology (ICSICT)* (2018), pp. 1–3
15. P. Maidee, C. Neely, A. Kaviani, C. Lavin, An open-source lightweight timing model for RapidWright. **12** 171–178 (2019)
16. M. Al-Qawasmi, A.G. Ye, An investigation of the accuracy of the VPR and COFFE area models in predicting the layout area of FPGA lookup tables, in *2020 SoutheastCon* (2020), pp. 1–9

17. Y. Pang, J. Xu, Z. Lu, Z. Li, Y. Zhang, J. Lai, Research on area modeling methodology for FPGA interconnect circuits, in *2019 IEEE 13th International Conference on ASIC (ASICON)* (2019), pp. 1–4
18. Synopsys, Liberty user guides and reference manual suite version 2017.06. https://www.academia.edu/43052430/Liberty_User_Guides_and_Reference_Manual_Suite_Version_2017_0620200514_69980_1frn721
19. IEEE standard for integrated circuit (IC) delay and power calculation system, in *IEEE Std. 1481-1999* (2000), pp. 1–400
20. IEEE standard for standard delay format (SDF) for the electronic design process, in *IEEE Std. 1497-2001* (2001), pp. 1–80

Chapter 3
Design (Application Design) Modeling

Abstract Application design is the bridge between end user's idea and FPGA's functional units. Modeling it will build up application design data structure—the ballast stone of any EDA engine in this stage. This chapter dives into the principles and implementations of FPGA design (application design) modeling, showing that how these models are classified and described.

3.1 Design Description Levels

3.1.1 Abstract Levels

FPGA application design's abstract levels are quite similar to CPU's: natural level [1], high level [2], low level [3], machine level [4] and physical level (Fig. 3.1).

In the field of CPU (scalar computing) application design, machine language is a series of instruction sequences composed of "0" and "1", directly interacting with the hardware at the bottom layer (for FPGA application design, the binary bitstream system); assembly language use abbreviated identifiers in its instructions to operate on the hardware (for FPGA application design, hardware description language is widely used to describe hardware circuits, requiring developers to have a considerable degree of low-level hardware knowledge); high-level language is more or less independent to a particular type of computing architecture and has already been the first choice for most computer programmers (for FPGA application design, it is also getting more and more commonly used); natural language is the way of communication between humans and considered to be the ultimate way to communicate with computing engines; there has been lots of research works in processing it (for FPGA application design, there is still a long way to go).

Just like chip design models (Sect. 2.1), application design models can also be theoretically described at similar abstract levels (Fig. 3.2). Related formats and standards have been intensively studied in this field to improve the design productivity.

K. Tu et al., *FPGA EDA*, https://doi.org/10.1007/978-981-99-7755-0_3

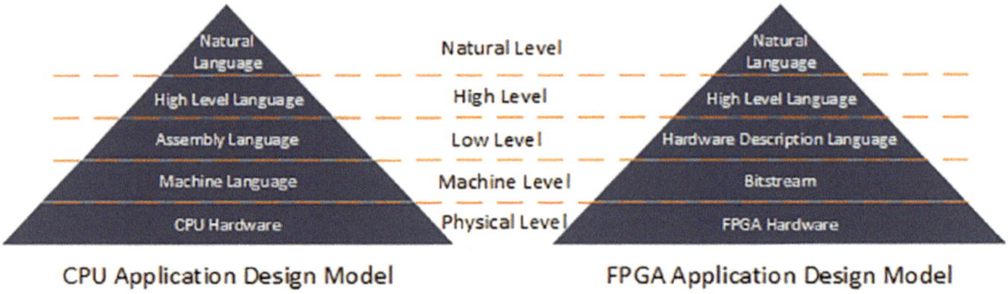

Fig. 3.1 Abstract levels of application design

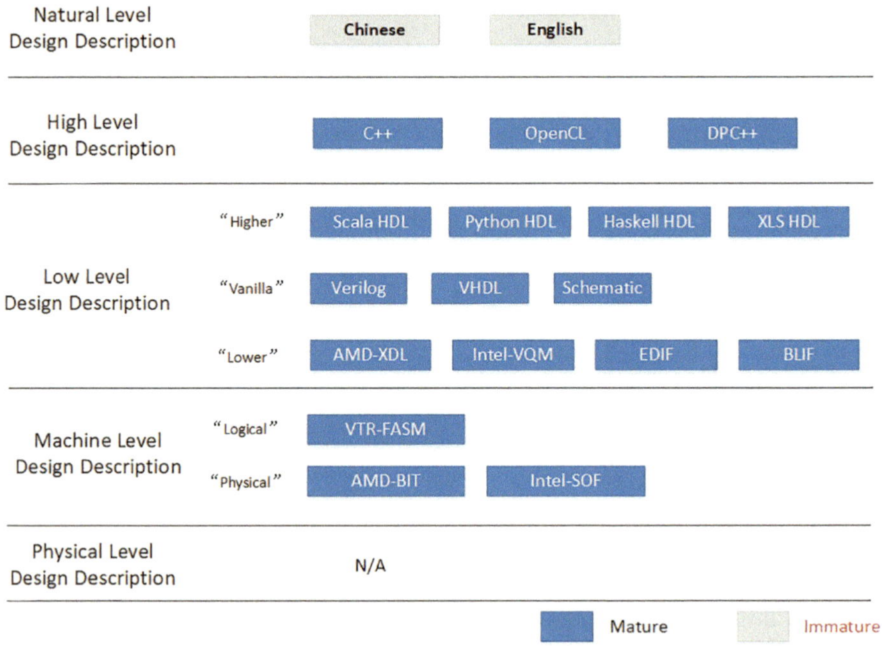

Fig. 3.2 FPGA design description abstract levels

1. Machine-Level Description
   The machine-level description of an application design refers to the expression of the bitstream (generally in binary), which directly controls the hardware behavior of the FPGA. Similar to device models, the "logical" description presents the correlation between every programmable bits in the bitstream and hardware resources (such as the FASM format file [5]); the "physical" description then is the final value of each configuration bits in a physical order determined by the configuration protocols (such as AMD's BIT format file and Intel's SOF format file).

2. Low-Level Description
   Identical to device models, three sub-levels are shown here:
   At "vanilla" level, traditional hardware description (such as Verilog, VHDL, or Schematics) is widely used to build the design model.
   At "higher" level, design descriptions with higher abstraction (such as Scala HDL, Python HDL, Haskell HDL, XLS HDL) could be a powerful complement to traditional descriptions.
   At "lower" level, more detailed hardware descriptions (such as AMD-XDL, Intel-VQM, BLIF, EDIF) is used to specify lower units in the FPGA design.
3. High-Level Description
   The high-level description for FPGA application design refers to software-oriented languages (such as C/C++, OpenCL, SystemC, DPC++).
   Inspired by the Open Computing Language (OpenCL) programming for heterogeneous systems, Intel has defined the Data Parallel C++ (DPC++) design language as its cross-architecture (CPU, GPU, FPGA) programming.
4. Natural-Level Description
   The automatic conversion of natural language into a language that FPGA can "understand" is also the future research direction of the academic community.

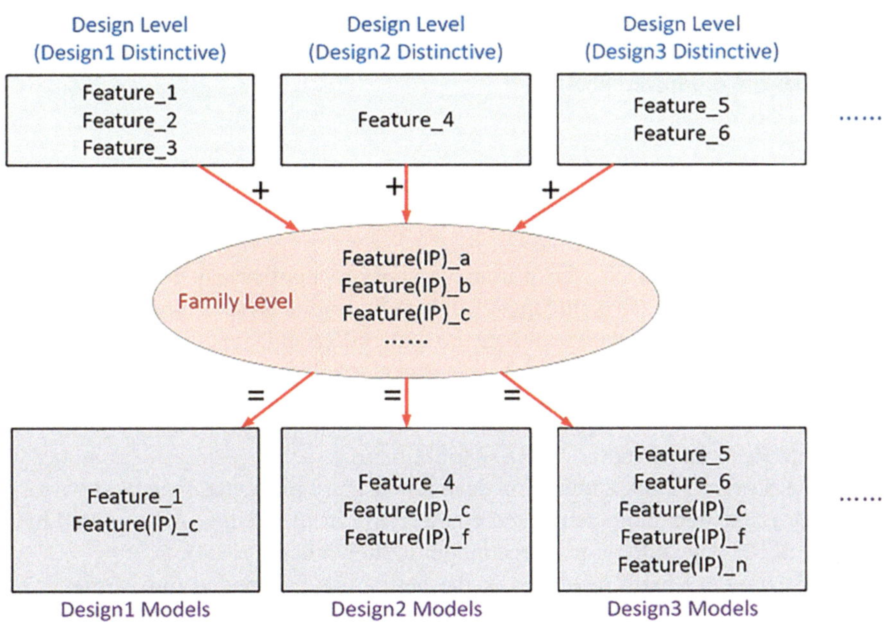

Fig. 3.3 FPGA design description reuse levels

3.1.2 *Reuse Levels*

From application design perspective, IP-based design methodology is the mainstream way of increasing reusability. IP (both soft ones and hard ones embedded in the FPGA) is generally family shared, which means it can be called when using any device under the supported families (Fig. 3.3). In modern FPGA application design EDA tools, IP integrator is an standard function that will not be absent, enabling users to get fast access to these predefined units.

3.2 Design Model Classifications

Similar to device (chip design) information, the design (application design) information of an FPGA can also be organized in classes: primary class, constraint class and report class. The "design checkpoint", built from these models, contains all the EDA information related to the application design.

The primary information is the torso of the design, the constraint information is set to direct the working strategy of EDA engines, then the report information shows the concerned metrics, helping designers to better analyze the current situation. If the reported results are not satisfactory, the design will be modified and then recurrently approaches the optimized goal.

3.2.1 *Primary Class*

Identical to device models, the primary models of application design EDA also include logical resource structure model and configuration bit structure model. Nevertheless, the substantial contents of them are quite different from the previous chapter. Again, the same with device models, we introduce design primary models at low abstract level (Fig. 3.1) for the same reason.

1. Logical Resource Structure (LRS) Model
   The LRS of an FPGA application design is usually presented by netlist—a term that describes the components and connectivity of the design. A simplified hierarchy of the design logic resource model is shown in (Fig. 3.4).
   The design core logic resources in the netlist can be divided into *clusters*(will accommodate in *tiles* in the device logic resources), each *cluster* is composed of *molecules*(will accommodate in *sites* in the device logic resources), and each *molecule* is composed of *atoms*(will accommodate in *primitives* in the device logic resource). Similarly, *atom* is also composed of gate-level units.
   The design interconnect logic resources in the netlist is composed of *nets*, and a *net* represents the connections between FPGA units (the edges of the netlist

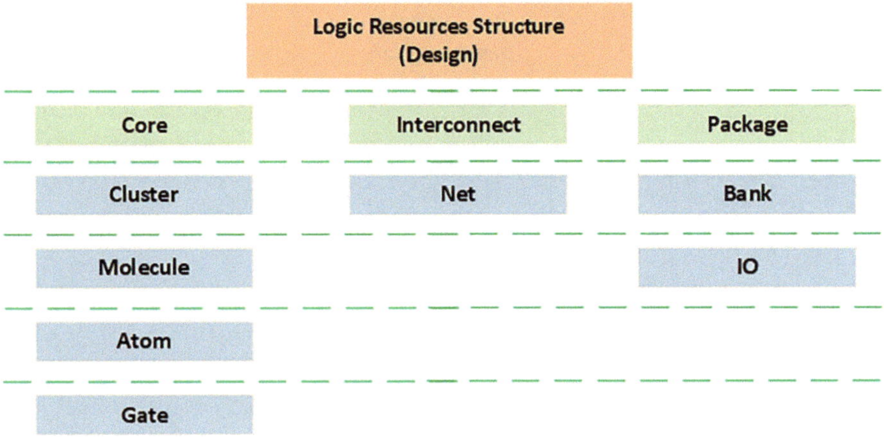

Fig. 3.4 FPGA design logical resource structure level

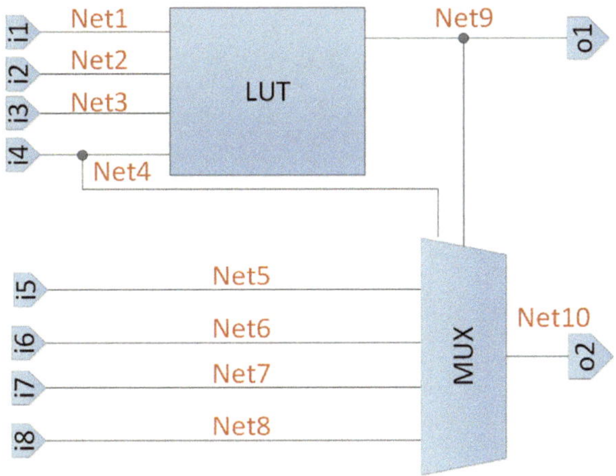

Fig. 3.5 FPGA application design netlist example

hyper-graph). Each net has a single driver pin, and a set of sink pins (will accommodate in *wires/switches* in the device logic resource).
The design IO will accordingly accommodate in Pad units in the device logic resource.
Take (Fig. 3.5) as an example, there are 12 *atoms*(1 LUT, 1 MUX, 8 inputs and 2 outputs) and 10 *nets* joining them altogether.

2. Configuration Bit Structure (CBS) Model
The CBS of an FPGA application design can also be defined from two perspectives: logical and physical.
Logical bit structure collects every active configuration bit's "logical address" of the design, that is, which logic resource it belongs to (Fig. 3.6).

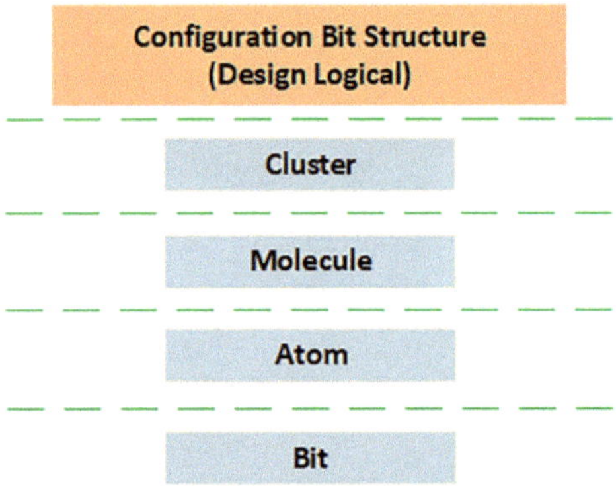

Fig. 3.6 FPGA design configuration bit structure level (logical)

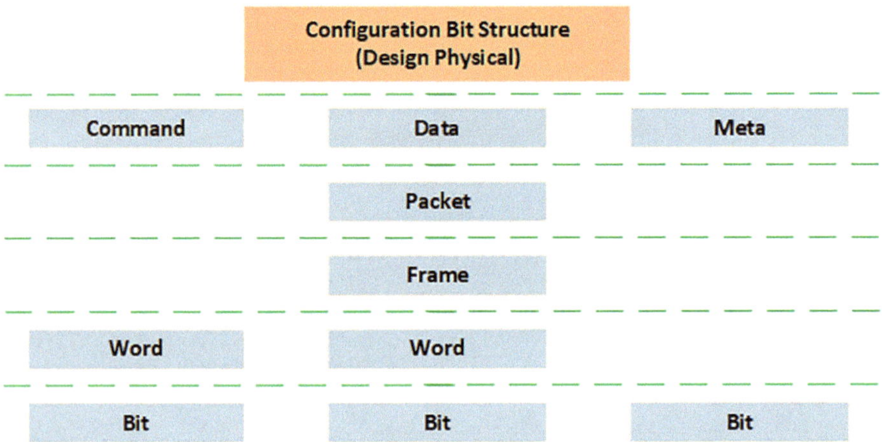

Fig. 3.7 FPGA design configuration bit structure level (physical)

Physical bit structure collects every active configuration bit's "physical address", that is, which position it lies in the final bitstream according to the programming protocol (Fig. 3.7).

After the logical and physical structure are properly identified, the configuration data can be outputted as the desired bitstream format (Fig. 3.8).

3.2.2 Constraint Class

FPGA application design constraints work at specific stage of the design flow, for example, routing constraints are used during the routing stage. Over-constraining or under-constraining the design both may cause sign-off difficulties.

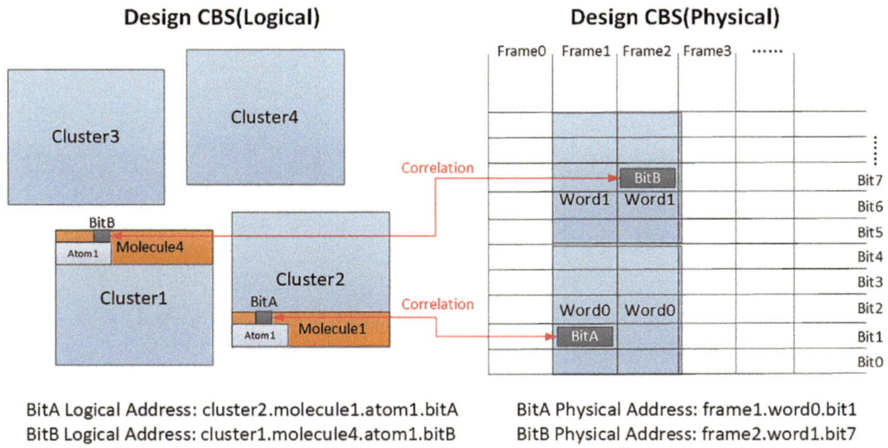

Fig. 3.8 FPGA design configuration bit correlation

TCL (Tool Command Language), pronounced "tickle", is an easy-to-learn scripting language and can run by scripts from either the Windows or Linux command-line. The language is easily extended with new function calls and has been expanded to support new tools and technology since its inception and adoption in the early 1990s. It has been adopted as the standard application programming interface, or API, among most EDA vendors to control and extend their applications.

Most of the FPGA vendors have adopted TCL as the design constraint format for their application EDA tools, as it is easily mastered by designers who are familiar with this industry standard language. The TCL interpreter inside the tool provides the full power and flexibility of TCL to control the flow or set the constraints.

Modern FPGA application design constraints have the following properties:

1. Inherit from industry standard SDC (Synopsys Design Constraint) commands and have its own expansions.
2. They are not simple strings, but are commands that follow the TCL semantic.
3. They can be interpreted like any other TCL command by the TCL interpreter.
4. They are read in and parsed sequentially the same as other TCL commands.

3.2.3 Report Class

Based on the objective (or EDA process) it addressed, the design reports can be divided into many categories: high-level synthesis report, logic synthesis report, physical implementation (packing/placement/routing...) report, analysis (timing/power/resource...) report, bitstream configuration (generation/download) report, and so on.

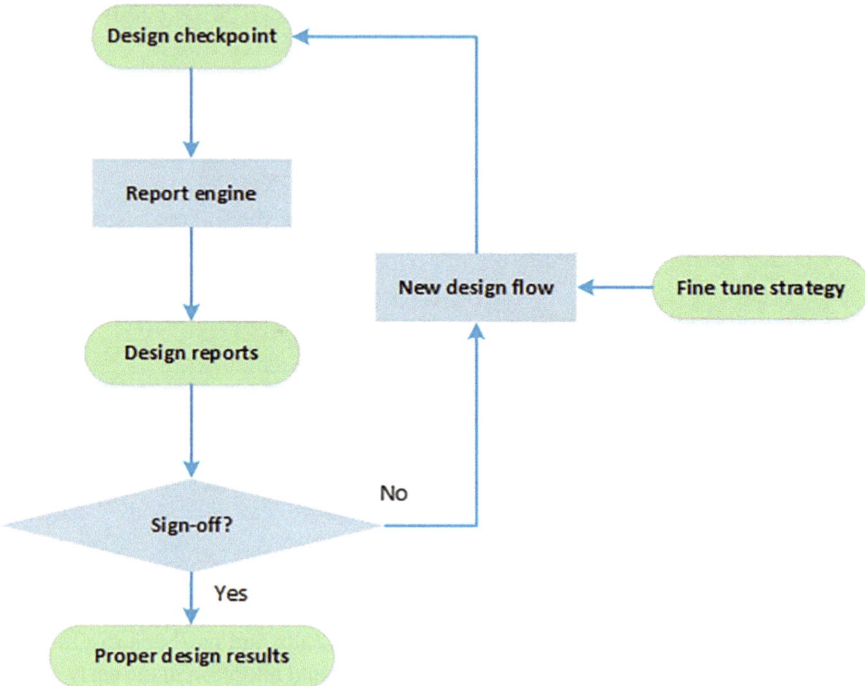

Fig. 3.9 FPGA application design report helps designers to sign-off properly

Design report offers information in human readable format from a specific perspective to help designers focus on the metrics they concern, if any sign-off requirement is not met, iterative modifications can be done until getting the proper solution (Fig. 3.9).

3.3 Design Model Implementations

The previous section listed all the design model classes: primary class, constraint class, and report class. In this section, we will present typical implementation practices of each model (Table 3.1).

3.3.1 Logic Resource Structure Model

In FPGA application design flow, the design netlist carries different information at different EDA stages. At logic synthesis stage, elaboration process turn the design

Table 3.1 Comparison of FPGA design model implementations

Model name	Abstract level	Reuse level	Class
Logic Resource Structure	High/Low	Design	Primary
Configuration Bit Structure	Machine	Design	Primary
Constraint	High/Low	Design	Constraint
Report	High/Low	Design	Report

Table 3.2 FPGA application design netlist formats and the EDA information they could carry ([a] is closed source)

Format	Generic netlist	Synthesized netlist	Implemented netlist	Adopter
RTLIL	Yes	/	/	Yosys
BLIF	Yes	Yes	/	Academia
GTECH[a]	Yes	/	/	Synplify
EDIF	Yes	Yes	/	Industry
VQM	/	Yes	/	Quartus
XDL	/	Yes	Yes	ISE
XDEF[a]	/	Yes	Yes	Vivado
VPR-Verilog	/	/	Yes	VPR
F4PGA-JASON	/	Yes	Yes	NextPnR

into gate-level representation (Generic Netlist), mapping process turn the design into atom-level representation (Synthesized Netlist); at physical implementation stage, cluster-level representation (Implemented Netlist) is generated.

There is no universal FPGA netlist format that can be used throughout the whole EDA process by the time this book is written, however, (Table 3.2) still listed the most popular legacy netlist formats and the EDA stages they could go through.

Implementation example: BLIF [6]

Berkeley Logic Interchange Format (BLIF) aimed to describe a logic-level hierarchical circuit in textual form.

Implementation example: EDIF [7]

Electronic Design Interchange Format (EDIF) is a format that could capture all features of circuit design. It has been accepted as a communications medium to manufacturing equipment and an interchange format between EDA systems.

Implementation example: Intel/Altera VQM [8]

Verilog Quartus Mapping (VQM) is the Intel/Altera version text file that contains a atom-level netlist. VQM files are typically generated by Intel/Altera Quartus.

Implementation example: AMD/Xilinx XDL [9, 10]

Xilinx Design Language (XDL) is the AMD/Xilinx version text file that represents a design netlist after mapping to the FPGA primitives. XDL files are typically generated by AMD/Xilinx ISE.

Table 3.3 SDC Syntax

Command	Supported arguments
Mostly [Verb]_[Noun]	Object / [-argument object]

```
<!-- Post-synthesis design report in XML -->
<atom_usage_report>
  <atoms num="<int>">
    <atom type="<atom_type_name_0>" usage="<int>"></atom>
    <atom type="<atom_type_name_1>" usage="<int>"></atom>
    ......
    <atom type="<atom_type_name_n>" usage="<int>"></atom>
  <input_pins num="<int>"></input_pins>
  <output_pins num="<int>"></output_pins>
  <nets num="<int>"></nets>
</atom_usage_report>
```

Fig. 3.10 Example XML syntax for post-synthesis design report

3.3.2 Configuration Bit Structure Model

1. Logical CBS information
   Implementation example: VTR-FASM [11]
   FPGA Assembly (FASM) is a textual representation of a bitstream. By assigning a symbolic name to each configurable thing in the FPGA, the resulting FASM file shows what features are specifically configured "on". These files provide an easy way to write programs that manipulate bitstreams. Modifying a textual FASM file is far easier than trying to modify a binary bitstream.
2. Physical CBS information
   Implementation example: AMD/Xilinx-BIT [12, 13]
   BIT files are AMD/Xilinx FPGA configuration files containing configuration information. In this file, each four bytes is a packet (analogous to CPU instruction). The packet could be a special header, or only carrying normal data. The header packet follows a simple assembly-like instruction set to dictate the configuration process.

3.3.3 Constraint Model

Synopsys's design constraint model (SDC) (Table 3.3) is the heart of all modern FPGA application design constraint models.

Implementation example: xDC ("x" represents the vendor)

FPGA vendors usually extend their constraint syntax based on SDC (because SDC cannot cover some FPGA specific syntax, such as physical constraints).

```
<!-- Packing report in XML -->
<packing_report>
  <cluster num="<int>">
    <unit name="<cluster_name_0>" accommodation_type="tile_type0"/>
    <unit name="<cluster_name_1>" accommodation_type="tile_type1"/
    ......
  </cluster>
  <input_pins num="<int>"/>
  <output_pins num="<int>"/>
  <nets num="<int>"/>
</packing_report>
```

Fig. 3.11 Example XML syntax for packing report

```
<!-- Placement report in XML -->
<placement_report>
  <unit name="cluster_name_0" accommodation_address="x0_y0_z0"/>
  <unit name="cluster_name_1" accommodation_address="x1_y1_z1"/>
  ......
</placement_report>
```

Fig. 3.12 Example XML syntax for placement report

```
<!-- Routing report in XML -->
<routing_report>
  <net name="net_name_0">
    <unit start="pin_0" end="pin_1"/>
    <unit start="pin_2" end="pin_3"/>
    <branch>
      <unit start="pin_0" end="pin_1"/>
      <unit start="pin_2" end="pin_3"/>
    </branch>
    <branch/>
    ......
  </net>
  <net name="net_name_1"/>
  ......
</routing_report>
```

Fig. 3.13 Example XML syntax for routing report

Universal FPGA constraint syntax still needs time to be standardized across vendors.

3.3.4 Report Model

Each FPGA vendor or academic organization has its own reporting style. Universal FPGA report syntax still needs time to emerge.

```
<!-- Power report in XML -->
<power_report>
  <power_models name="final"/>
  <total_power value="<float>"/>
  <transceiver_static_power value="<float>"/>
  <transceiver_dynamic_power value="<float>"/>
  <core_static_power value="<float>"/>
  <core_dynamic_power value="<float>"/>
  ......
</power_report>
```

Fig. 3.14 Example XML syntax for power report

```
<!-- Timing report in XML -->
<timing_report>
  <path start="atom_0" end="atom_1" type="setup/hold"
slack="<float>"/>
  <path start="atom_2" end="atom_3" type="setup/hold"
slack="<float>"/>
  ......
</timing_report>
```

Fig. 3.15 Example XML syntax for timing report

1. Post-synthesis report
   Implementation example: (Fig. 3.10)
2. Post-implementation report
   Implementation example: (Figs. 3.11, 3.12 and 3.13)
3. Power report
   Implementation example: (Fig. 3.14)
4. Timing report
   Implementation example: (Fig. 3.15)

References

1. Wikipedia, Natural language (2022). https://en.wikipedia.org/wiki/Natural_language
2. Wikipedia, High level language (2022). https://en.wikipedia.org/wiki/High-level_programming_language
3. Wikipedia, Assembly language. (2022) https://en.wikipedia.org/wiki/Assembly_language
4. Wikipedia, Machine code (2022). https://en.wikipedia.org/wiki/Machine_code
5. F4PGA, FPGA assembly (FASM) (2021). https://fasm.readthedocs.io/en/latest/
6. U. of California Berkeley, Berkeley logic interchange format (1992). https://people.eecs.berkeley.edu/~alanmi/publications/other/blif.pdf
7. H.J. Kahn, R.F. Goldman, The electronic design interchange format EDIF: present and future, in *Proceedings of the 29th ACM/IEEE Design Automation Conference*, Series DAC '92 (IEEE Computer Society Press, Washington, DC, USA, 1992), pp. 666–671
8. A. QUIP, VQM extractor and language functional description (2005)

9. C. Beckhoff, D. Koch, J. Torresen, The Xilinx design language (XDL): tutorial and use cases, in *6th International Workshop on Reconfigurable Communication-Centric Systems-on-Chip (ReCoSoC)* (2011), pp. 1–8
10. B.J.P. Tomas, Xilinx design language (2012). http://www.ee.unlv.edu/~selvaraj/ecg707/lecture/XilinxDesignLanguage.pdf
11. B.J.P. Tomas, FPGA assembly (FASM). https://fasm.readthedocs.io/en/latest/
12. AMD/Xilinx, Xilinx bit bitstream files. http://www.pldtool.com/pdf/fmt_xilinxbit.pdf
13. Y. Shan, FPGA bitstream explained. http://lastweek.io/fpga/bitstream/

Part III
FPGA Metric Analysis

Chapter 4
Power Analysis

Abstract Power dissipation has become one of the top concern in the development of new integrated circuits. In this chapter, power analysis techniques for FPGA are introduced. The power consumption of an FPGA depends on both chip design and application design, and as the capacities of FPGAs continue to grow, the challenge of power efficiency will only increase.

4.1 Overview

For any electronic device, power is an eternal topic that must be faced. In most cases, FPGA consumes much more power than their ASIC counterparts because extra resources are utilized to ensure its programmability [1].

Figure 4.1 gives the composition of power dissipation of general CMOS circuits, and it is also suitable for FPGAs.

Accordingly, the total power usage of an FPGA device (P_{Total}) can be broken down as Eq. 4.1.

$$P_{\text{Total}} = P_{\text{Static}} + P_{\text{Dynamic}} = P_{\text{switching}} + P_{\text{short-circuit}} + P_{\text{leakage}} \tag{4.1}$$

FPGA static power is the transistor leakage power (pleakage) when the device is powered and and configured. It is proportional to the static current—the current that flows regardless of gate switching (transistor is ON "biased" or OFF "unbiased"). As technology advances, this power is becoming non negligible due to the shrinking of transistors' size as well as the thickness of the oxides.

FPGA dynamic power is the additional power consumption caused by application design's signal activities. Switching power ($P_{\text{switching}}$) is dissipated when charging or discharging internal and net capacitance, short-circuit power ($P_{\text{short-circuit}}$) is the power dissipated by an instantaneous short-circuit connection between the supply voltage and the ground at the time the gate switches state.

K. Tu et al., *FPGA EDA*, https://doi.org/10.1007/978-981-99-7755-0_4

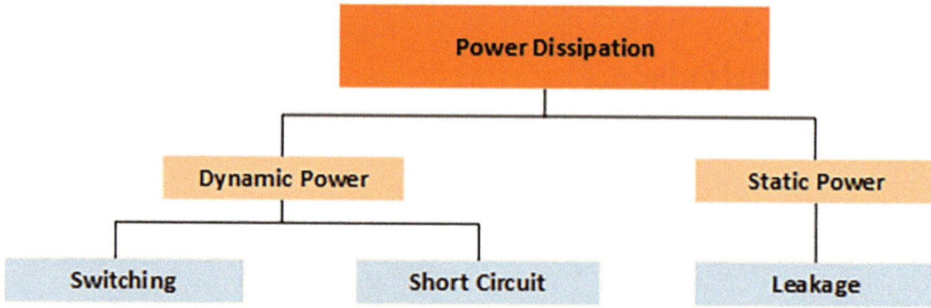

Fig. 4.1 Power dissipation type of CMOS circuits

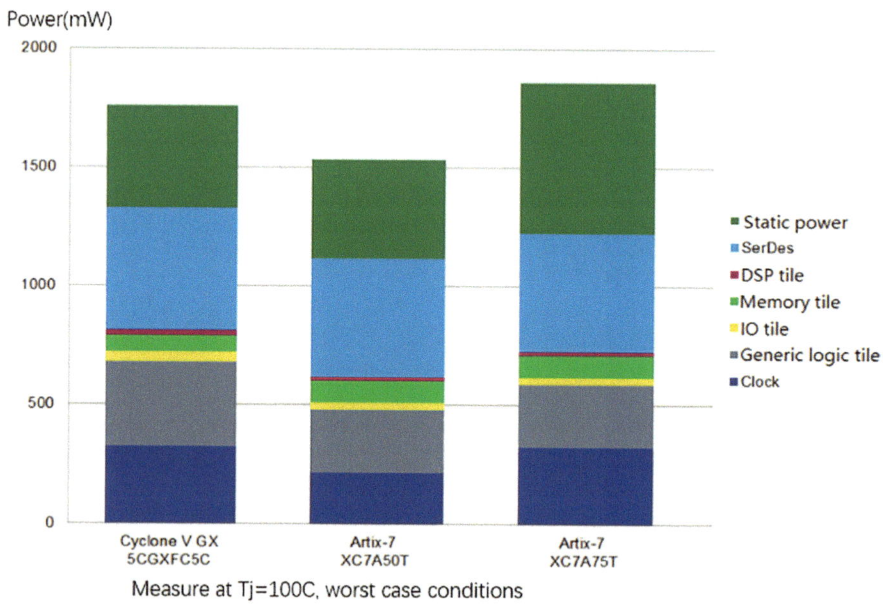

Fig. 4.2 Power dissipation of some classic FPGAs. *Source* Microchip/Microsemi

Researchers have investigated the power dissipation for each part of an FPGA [2, 3], and apart from academia, power distribution is also one of the top concerns in industry (Fig. 4.2).

Power analysis engines use basic theories (Eqs. 4.2–4.4) to calculate the power consumption for every FPGA unit and add them up together to get the final result. Power analysis engines can run at different abstract levels (introduced in Sect. 3.1.1), such as high level [4–6],

$$P_{\text{switching}} = \alpha \cdot C \cdot V_{\text{dd}}^2 \cdot f \tag{4.2}$$

where α = switching activity factor, C = the effective capacitance, V_{dd} = the supply voltage, and f = switching frequency.

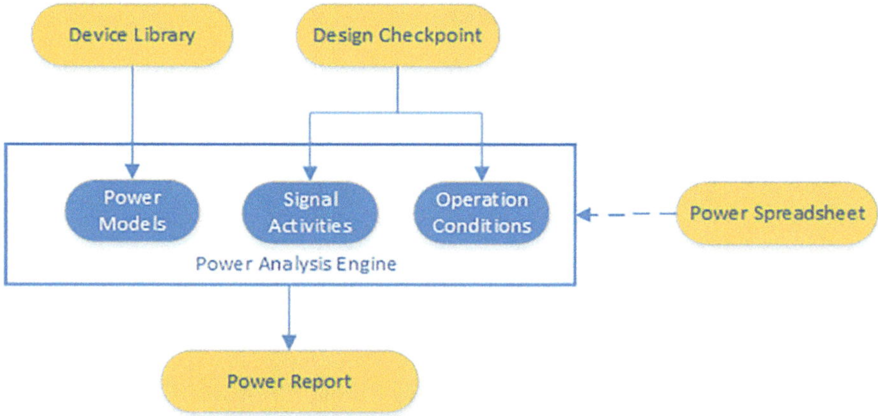

Fig. 4.3 Typical inputs and outputs in power analysis flow

$$P_{\text{short-circuit}} = I_{sc} \cdot V_{dd} \cdot f \tag{4.3}$$

where I_{sc} = the short-circuit current during switching, V_{dd} = the supply voltage and f = switching frequency.

$$P_{\text{leakage}} = V_{dd} \cdot I_{\text{leakage}} \tag{4.4}$$

where V_{dd} = the supply voltage, I_{leakage} = the leakage current.

As Fig. 4.3 illustrates, the target design checkpoint (containing signal activities and operation condition constraints) and device library (containing power models) are the main inputs that a power analysis engine generally needs. Most FPGA companies provide power spreadsheets for their customers to estimate power dissipation for particular devices [7, 8]. Instead of design checkpoint and device library previously mentioned, these manual spreadsheets basically covered all the information that a power estimator needs and could be very useful at the early stage of the design. The final output is a power report showing the power details of FPGA.

Given that device power model has been discussed in Sect. 2.2.2, design operation conditions (such as temperature, voltage, process) can be easily obtained by constraints, the remaining pillar of the engine—signal activities—is coming under the spotlight.

Different techniques can be applied to get signal activities in an FPGA [9]—simulation-based and probabilistic-based. Simulation-based techniques are used by most of the EDA tools due to their precision and generosity [10, 11]. They simulate the target design for a predetermined operation condition by applying data stimuli to the inputs. Sometimes, these input patterns can be randomly generated and simulation results are calculated statistically until a desired precision is achieved [12–15]. On the other hand, probabilistic-based techniques use input characteristics instead of the real input vectors. They generally rely on the static probability and the transition probability of each signal [2, 16, 17].

4.2 Power Analysis Techniques

Both dynamic and static power estimations are dependent on the behavior of the application design. These designs usually do not fully utilize the available programmable resources in the FPGA. The dynamic power will vary according to the resource utilization and algorithmic function of the application design.

The QoR (quality of results) of power analysis is strongly influenced by the quality of the signal activity data.

1. Simulation-based
   For simulation-based techniques, it often performs at lower level of the circuit design and results in better accuracy. One of the most popular simulation tools is at transistor level—Simulation Program with IC Emphasis (SPICE).
   The general analysis technique can be summarized as follows [18]:

   a. Run logic simulation with a set of input vectors.
   b. Monitor the switching activity of each unit (depends on the simulation level), then calculate the dynamic and static power using equations (e.g., Eqs. 4.2–4.4).
   c. Determine the total power dissipation by summing the dynamic and static power obtained in previous step.

   Simulation-based technique is widely used in industry. Both AMD and Intel's FPGA power tools calculate signal activities from dedicated simulation output files, such as Value Change Dump(VCD) or Switching Activity Interchange Format (SAIF). There is a great deal of information present in VCD waveform files, and this can be reduced to a smaller dataset for average power estimation and for power optimization. The SAIF file format is used for this purpose.

   a. VCD
      VCD file contains value changes of signals, for example, at what times signals changes their values. VCD format is part of the IEEE standard for Verilog (discussed starting on page 325) [19].
   b. SAIF
      SAIF file contains toggle counts and time information like how much time a signal was in 1 state, 0 state, or x state.

   VCD file can be converted into SAIF file using Synopsys vcd2saif tool, because effectively VCD file is a superset of SAIF file. Power estimation with time stamps of individual value changes must use VCD file. If you have a VCD file which is simulated between 0 and 10ns, then you can do power analysis for any time range between 0 and 10ns, for example, power for 2-5ns, 3-7ns...but for SAIF file, you can not do the same thing. You have to regenerate SAIF files for these time intervals.
2. Probabilistic-based
   For probabilistic-based techniques, the behavior of the inputs can be characterized by parameters:

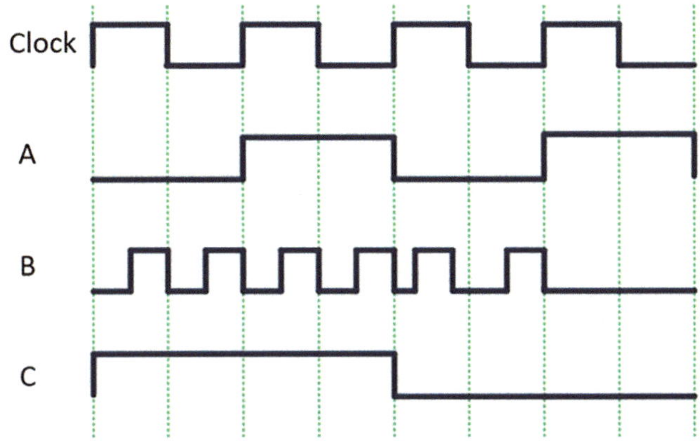

Signal	Static Probability	Transition Density
Clock	2/4=0.500	8/4=2.0
A	2/4=0.500	4/4=1.0
B	1.5/4=0.375	12/4=3.0
C	2/4=0.500	2/4=0.5

Fig. 4.4 Examples of signals represented static probabilities and transition densities

a. The static probability, P_1 P_1 is the long-term probability that a signal is logic high. For example, a clock signal with a 50% duty cycle will have $P_1(\text{clk}) = 0.5$.
b. The transition density, T_d T_d is the average number of times the signal will switch during each clock cycle. For example, a clock signal with a 50% duty cycle will have $T_d(\text{clk}) = 2.0$.

Figure 4.4 shows examples of how these parameters are defined.

After these parameters are known for all FPGA units, power estimation can then do the calculation. For example, dynamic power is directly proportional to A_s, and static power is dependent on P_1.

3. Simulation/Probabilistic hybrid
   There are many academic exploration in simulation/probabilistic hybrid methods. ACE-2.0 algorithm [20] is the representative one, and it uses both simulation and static analyses to determine the parameters (P_1, A_s) that probabilistic-based techniques would use.

4.3 Summary and Trends

In this section, we dived into the composition of power consumption of FPGA and analyzed each component in theory. Except for power-related information in FPGA device model, signal activities derived from FPGA design model/checkpoint is another important data source for power estimation.

Simulation-based (vector dependent) and probabilistic-based (vector independent) techniques are the most popular approaches to achieve signal activities. However, pros and cons of these two methods are obvious: simulation-based techniques have high accuracy and generality but consumes more time and memory resources, whilst probabilistic-based techniques are quite opposite—better time efficiency but less accuracy. To trade off estimation speed and accuracy, a hybrid of the two techniques is the way relatively easy to think of.

With the increasing complexity of FPGA designs and the need for low-power designs, several trends have emerged in FPGA power analysis:

1. Power analysis using AI (machine learning)
   Machine learning techniques are being used to optimize the power consumption of FPGA designs. These techniques analyze the design and provide recommendations to optimize the power consumption.
2. Power analysis at higher levels of abstraction
   Power analysis is being performed at higher levels of abstraction such as system-level and behavioral level. This approach enables early power estimation and optimization, reducing the overall design time.
3. Power analysis for heterogeneous architectures
   FPGA designs are increasingly using heterogeneous architectures such as CPU-FPGA, GPU-FPGA, and ASIC-FPGA. Power analysis for these architectures is challenging and requires the development of new power analysis techniques.
4. Power analysis for security
   Power analysis is being used for security purposes such as side-channel analysis and fault injection analysis. These techniques analyze the power consumption of the FPGA to detect security vulnerabilities.

References

1. I. Kuon J. Rose, Measuring the gap between FPGAs and ASICs, *IEEE Transactions on Computer-Aided Design of Integrated Circuits and Systems*, vol. 26, no. 2 (2007), pp. 203–215
2. T. Osmulski, J.T. Muehring, B. Veale, J.M. West, H. Li, S. Vanichayobon, S.-H. Ko, J.K. Antonio, S.K. Dhall, A probabilistic power prediction tool for the Xilinx 4000-series FPGA, in *Parallel and Distributed Processing*, ed. by J. Rolim (Springer, Berlin, Heidelberg, 2000), pp. 776–783
3. L. Shang, A.S. Kaviani, K. Bathala, Dynamic power consumption in Virtex^{TM}-II FPGA family, in *Proceedings of the 2002 ACM/SIGDA Tenth International Symposium on Field-Programmable Gate Arrays*, series FPGA '02 (Association for Computing Machinery, New York, NY, USA, 2002), pp. 157–164. Available https://doi.org/10.1145/503048.503072

4. D. Chen, J. Cong, Y. Fan, Z. Zhang, High-level power estimation and low-power design space exploration for FPGAs, in *2007 Asia and South Pacific Design Automation Conference* (2007), pp. 529–534
5. Z. Lin, Z. Yuan, J. Zhao, W. Zhang, H. Wang, Y. Tian, PowerGear: early-stage power estimation in FPGA HLS via heterogeneous edge-centric GNNs, in *2022 Design, Automation & Test in Europe Conference & Exhibition (DATE)* (2022), pp. 1341–1346
6. Z. Lin, T. Liang, J. Zhao, S. Sinha, W. Zhang, Hl-pow: learning-assisted pre-RTL power modeling and optimization for FPGA HLS. IEEE Trans. Comput.-Aided Des. Integr. Circuits Syst. 1–1 (2023)
7. Intel, Early power estimator user guide (2021). https://www.intel.com/programmable/technical-pdfs/683272.pdf
8. Xilinx, Xilinx power estimator user guide (2022). https://www.xilinx.com/support/documents/sw_manuals/xilinx2022_1/ug440-xilinx-power-estimator.pdf
9. J.B. Goeders, S.J.E. Wilton, VersaPower: Power estimation for diverse FPGA architectures, in *2012 International Conference on Field-Programmable Technology* (2012), pp. 229–234
10. X. Tang, P.-E. Gaillardon, G. De Micheli, FPGA-spice: a simulation-based power estimation framework for FPGAs, in *2015 33rd IEEE International Conference on Computer Design (ICCD)*, (2015), pp. 696–703
11. S. Seeley, V. Sankaranaryanan, Z. Deveau, P. Patros, K.B. Kent, Simulation-based circuit-activity estimation for FPGAs containing hard blocks, in *2017 International Symposium on Rapid System Prototyping (RSP)* (2017), pp. 36–42
12. N. Burch, Yang, Trick, McPOWER: a Monte Carlo approach to power estimation, in *1992 IEEE/ACM International Conference on Computer-Aided Design* (1992), pp. 90–97
13. E. Todorovich, E. Boemo, F. Angarita, J. Vails, Statistical power estimation for FPGAs, in *International Conference on Field Programmable Logic and Applications, 2005* (2005), pp. 515–518
14. Y.A. Durrani, T. Riesgo, Efficient power analysis approach and its application to system-on-chip design, Microprocess. Microsyst. **46**(PA):11–20 (2016). Available https://doi.org/10.1016/j.micpro.2016.09.003
15. G. Verma, C. Dabas, A. Goel, M. Kumar, V. Khare, Clustering based power optimization of digital circuits for FPGAs, J. Inf. Optim. Sci. **38**(6), 1029–1037 (2017)
16. F. Najm, A survey of power estimation techniques in VLSI circuits, IEEE Trans. Very Large Scale Integr. (VLSI) Syst. **2**(4), 446–455 (1994)
17. S. Garg, S. Tata, R. Arunachalam, Static transition probability analysis under uncertainty, in *IEEE International Conference on Computer Design: VLSI in Computers and Processors, 2004. ICCD 2004. Proceedings* (2004), pp. 380–386
18. G. K. Yeap, *Practical Low Power Digital VLSI Design* (Kluwer Academic Publishers, USA, 1998)
19. IEEE standard for Verilog hardware description language, in *IEEE Std. 1364-2005 (Revision of IEEE Std. 1364-2001)* (2006), pp. 1–590
20. J. Lamoureux, S.J. Wilton, Activity estimation for field-programmable gate arrays, in *2006 International Conference on Field Programmable Logic and Applications* (2006), pp. 1–8

Chapter 5
Performance (Timing) Analysis

Abstract Timing analysis can be static or dynamic. Dynamic timing analysis (DTA) verifies functionality of the design by applying input vectors and checking for correct output vectors whereas static timing analysis (STA) checks static delay requirements of the circuit without any input or output vectors. In this chapter, STA techniques is focused since it is widely used in FPGA design flow to make sure the timing requirements are met.

5.1 Overview

Dynamic timing analysis (DTA), also known as simulation-based timing analysis technique, is complicated for even small FPGAs because of huge number of input vectors and unbearable long simulation time, while static timing analysis (STA), which could analyze a design in a very short time, is then thriving. As a mainstay of modern FPGA design flows, STA breaks a design down into timing paths, calculates the signal propagation delay along each path, and checks for violations of timing constraints inside the design and at the input/output interface. STA also has been integrated with timing-driven EDA engines to optimize FPGA's timing performance.

The target design checkpoint (containing timing constraints and timing graph) and device library (containing timing models) are the main inputs that a timing analysis engine needs. The final output is the timing report (Fig. 5.1).

Standard Delay Format (SDF) is another optional output of timing engine. SDF is an IEEE standard for the representation and interpretation of timing data (both cell delays and interconnect delays) for use at any stage of the electronic design process [1]. This can be used along with the netlist in a simulator to verify that design meets its functional and timing requirements.

Given that device timing model has been discussed in Sect. 2.2.2, timing constraints has been discussed in Sect. 3.2.2, another main input–timing graph, derived from target design checkpoint, will be introduced in the following section.

Before we dive into the timing calculation algorithms, here are some basic concepts about STA. Figure 5.2 is the most common used picture to illustrate this.

K. Tu et al., *FPGA EDA*, https://doi.org/10.1007/978-981-99-7755-0_5

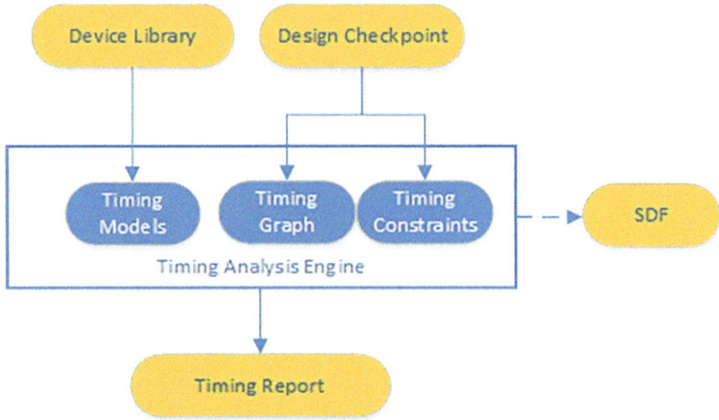

Fig. 5.1 Typical inputs and outputs in timing analysis flow

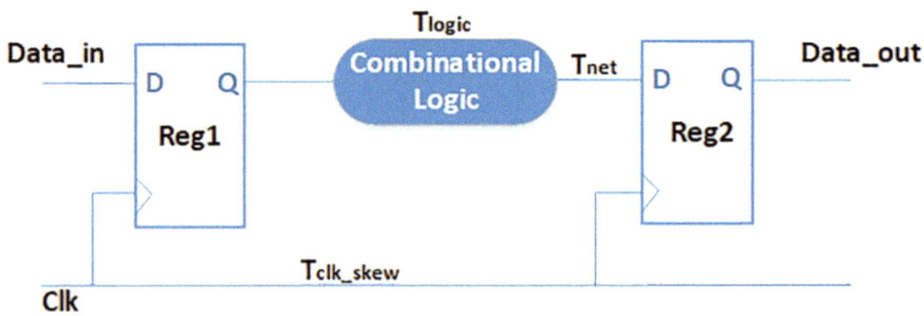

Fig. 5.2 Typical setup/hold timing analysis

Equations 5.1 and 5.2 can accurately represents the calculations of setup time slack ($\mathrm{Slack}_{\mathrm{setup}}$) and hold time slack ($\mathrm{Slack}_{\mathrm{hold}}$).

$$\mathrm{Slack}_{\mathrm{setup}} = T_{\mathrm{period}} - (T_{\mathrm{cq}} + T_{\mathrm{logic}} + T_{\mathrm{net}} + T_{\mathrm{setup}} - T_{\mathrm{clk}_{\mathrm{skew}}}) \tag{5.1}$$

$$\mathrm{Slack}_{\mathrm{hold}} = T_{\mathrm{cq}} + T_{\mathrm{logic}} + T_{\mathrm{net}} - T_{\mathrm{hold}} - T_{\mathrm{clk}_{\mathrm{skew}}}) \tag{5.2}$$

where T_{period} is clock period, T_{cq} is defined as time it takes for data to appear on output Q once clock is triggered (pos edge or neg edge), T_{logic} is the delay of the combinational logic, T_{net} is the delay of the routing net, $T_{\mathrm{clk}_{\mathrm{skew}}}$ is the time difference between the clock arriving time at the two flip-flops.

To simplify the equation, T_{net} and $T_{\mathrm{clk}_{\mathrm{skew}}}$ can be ignored. In order to make sure that $\mathrm{Slack}_{\mathrm{setup}}$ and $\mathrm{Slack}_{\mathrm{hold}}$ are positive, we can derive Eqs. 5.3 and 5.4 (plus T_{setup} on both sides) from Eqs. 5.1 and 5.2.

$$T_{\mathrm{period}} > T_{\mathrm{cq}} + T_{\mathrm{logic}} + T_{\mathrm{setup}} \tag{5.3}$$

$$T_{cq} + T_{logic} + T_{setup} > T_{hold} + T_{setup} \quad (5.4)$$

Combine Eqs. 5.1 and 5.2, we can have Eq. 5.5.

$$T_{hold} + T_{setup} < T_{cq} + T_{logic} + T_{setup} < T_{period} \quad (5.5)$$

$T_{cq} + T_{logic} + T_{setup}$ is the data propagation delay, if it is greater than T_{period}, the data will not arriving when the second register is sampling, on the other hand, if it is smaller than the register sampling window ($T_{hold} + T_{setup}$), the registers could fall into metastability.

In FPGA design, STA can be performed in different stages: post-synthesis (logical level) and post-implementation (physical level). Post-synthesis STA (based on ideal implementation information) is faster but less accurate than post-implementation STA (based on real implementation information).

5.2 Timing Analysis Techniques

STA usually requires a timing graph that describes the target design from the timing perspective, identifying all the timing paths. The timing graph consists of nodes and edges, nodes correspond to component pins or input/output ports, and edges are the timing path between them. Edges have attached weights that can denote some characteristics such as delay values [2].

Timing Graph Definition: A timing graph G = N, E, s, t is a directed graph having exactly one source node s and one sink node t , where N is a set of nodes, and E is a set of edges. The weight associated with an edge corresponds to either the gate delay or the interconnect delay (Fig. 5.3).

Traditional STA is deterministic (DSTA) and compute the circuit delay for a specific condition. In practice, the worst-case slow or best-case fast process is typically used and this could lead to over-design, leaving a lot of margin on the table in terms of PPA. Statistical STA (SSTA) then come out to address this problem. It combines

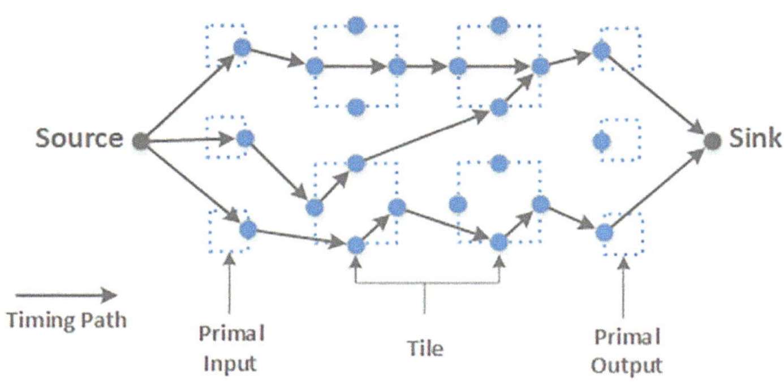

Fig. 5.3 Example of timing graph

the delays along the timing paths which is expressed statistically (with mean and standard deviations) to obtain the overall delay data.

SSTA is also employed by Intel in its Quartus Prime software to mitigate the effect of random variation on longer paths [3]. By discounting the minimum/maximum delay spread on these paths, the FPGA performance reported by STA may increase. There are two main categories of SSTA techniques–path-based and block-based.

1. Path-based
   In path-based STA technique, critical path is searched in an exhaustive way. The statistical calculation is simple, but the paths of interest must be identified prior to running the analysis [4–6].
2. Block-based
   In block-based STA technique, the circuit timing graph is traversed in a topological manner. In [7], two basic graph traversal algorithms–depth first search (DFS) and breadth first search (BFS) are applied to STA module and the runtime efficiencies is compared by testing a large number of sequential circuit instances. The conclusion is that BFS algorithm can implement STA module more efficiently than DFS algorithm. Due to its runtime advantage, many research [8–11] and commercial efforts have taken the block-based approach. The advantage is completeness, and no need for path selection, however, to compute statistical max (or min) of random variables is not trivial.

The choice of using path-based analysis or block-based analysis depends on several factors, such as the design complexity, stage, and goal. Generally, path-based analysis is more suitable for small or medium-sized designs, where the number of paths is manageable and the accuracy is important. It can also be used for final verification or optimization, where the timing margins are tight and the details are needed. On the other hand, block-based analysis is more suitable for large designs, where the number of paths is overwhelming and the runtime is important. It can also be used for FPGA architecture exploration, where the timing budget is loose and the trends are sufficient.

In some cases, it could be more optimized to combine both techniques and use them in different stages or levels of the design. For example, one can use block-based analysis for the system-level design, where the blocks are abstracted and the overall timing is estimated. Then, one can use path-based analysis for the block-level design, where the paths are detailed. The balance between accuracy and efficiency can be obtained in this way [12].

5.3 Summary and Trends

The state-of-the-art STA engines still can not replace DTA (simulation) completely because there are some aspects of timing verification that cannot yet be completely captured and verified in STA [13]. Some of these limitations include:

1. Inaccurate timing models
   The timing models used in FPGA STA may not accurately represent the behavior of the actual circuit due to the complexity of the FPGA architecture.
2. Lack of support for dynamic circuits
   FPGA STA assumes that the circuit is static and does not take into account dynamic circuits such as state machines or circuits with feedback paths.
3. Impact of environmental conditions
   FPGA STA assumes ideal environmental conditions, such as constant temperature and voltage, which may not hold true in the real world.

Although FPGA STA has been matured for many years, it still benefits from emerging technologies. The following are some of the recent trends in FPGA STA:

1. Parallel acceleration
   Parallel STA on different computing platforms is one of the researching hot spots, such as multi-core CPUs [4, 14–17] and GPUs [16, 18].
2. AI (machine learning) acceleration
   ML algorithms are increasingly being used to analyze the timing characteristics of FPGA designs [19–21]. ML-based timing analysis can quickly identify critical paths in the design, predict the timing behavior of the design, and optimize the design for timing performance.

References

1. IEEE standard for standard delay format (SDF) for the electronic design process, in *IEEE Std. 1497-2001* (2001), pp. 1–80
2. J.L.M. Lee, A scalable method to measure similarity between two EDA-generated timing graphs, in *2015 International Conference on Computer, Communications, and Control Technology (I4CT)* (2015), pp. 44–48
3. Intel, Guaranteeing silicon performance with FPGA timing models. https://cdrdv2-public.intel.com/650314/wp-01139-timing-model.pdf
4. T.-W. Huang, M.D.F. Wong, UI-timer 1.0: an ultrafast path-based timing analysis algorithm for CPPR. IEEE Trans. Comput.-Aided Des. Integr. Circuits Syst. **35**(11), pp. 1862–1875 (2016)
5. D. Mishagli, E. Koskin, E. Blokhina, Path-based statistical static timing analysis for large integrated circuits in a weak correlation approximation, in *2019 IEEE International Symposium on Circuits and Systems (ISCAS)* (2019), pp. 1–5
6. L.-W. Chen, Y.-N. Sui, T.-C. Lee, Y.-L. Li, M.C.-T. Chao, I.-C. Tsai, T.-W. Kung, E.-C. Liu, Y.-C. Chang, Path-based pre-routing timing prediction for modern very large-scale integration designs, in *2022 23rd International Symposium on Quality Electronic Design (ISQED)* (2022), pp. 1–6
7. J. Lu, N. Xu, J. Yu, T. Weng, Research of timing graph traversal algorithm in static timing analysis based on FPGA, in *2017 IEEE 3rd Information Technology and Mechatronics Engineering Conference (ITOEC)* (2017), pp. 334–338
8. L. Zhang, Y. Hu, C.-P. Chen, Block based statistical timing analysis with extended canonical timing model, in *Proceedings of the ASP-DAC 2005. Asia and South Pacific Design Automation Conference, 2005.*, vol. 1 (2005), pp. 250–253

9. R. Chen, H. Zhou, New block-based statistical timing analysis approaches without moment matching, in *Proceedings of the ASP-DAC 2007—Asia and South Pacific Design Automation Conference 2007*, Series. Proceedings of the Asia and South Pacific Design Automation Conference, ASP-DAC (2007), pp. 462–467
10. G. Luo, B. Jin, W. Zhang, A fast and simple block-based approach for common path pessimism removal in static timing analysis, in *2015 14th International Conference on Computer-Aided Design and Computer Graphics (CAD/Graphics)* (2015), pp. 234–235
11. L. Jin, W. Fu, Y. Zheng, H. Yan, A precise block-based statistical timing analysis with max approximation using multivariate adaptive regression splines, in *2019 IEEE 13th International Conference on ASIC (ASICON)* (2019), pp. 1–4
12. T.-W. Huang, M.D.F. Wong, OpenTimer: a high-performance timing analysis tool, in *2015 IEEE/ACM International Conference on Computer-Aided Design (ICCAD)* (2015), pp. 895–902
13. J. Bhasker, R. Chadha, *Static Timing Analysis for Nanometer Designs: A Practical Approach*, 1st edn. (Springer, 2009)
14. Y.-M. Yang, Y.-W. Chang, I.H.-R. Jiang, iTimerC: common path pessimism removal using effective reduction methods, in *2014 IEEE/ACM International Conference on Computer-Aided Design (ICCAD)* (2014), pp. 600–605
15. T.-W. Huang, M.D.F. Wong, D. Sinha, K. Kalafala, N. Venkateswaran, A distributed timing analysis framework for large designs, in *Proceedings of the 53rd Annual Design Automation Conference*, Series DAC '16 (Association for Computing Machinery, New York, NY, USA, 2016). Available https://doi.org/10.1145/2897937.2897959
16. K. E. Murray, V. Betz, Tatum: parallel timing analysis for faster design cycles and improved optimization, in *2018 International Conference on Field-Programmable Technology (FPT)* (2018), pp. 110–117
17. T.-W. Huang, G. Guo, C.-X. Lin, M.D.F. Wong, OpenTimer v2: a new parallel incremental timing analysis engine. IEEE Trans. Comput.-Aided Des. Integr. Circ. Syst. **40**(4):776–789 (2021)
18. Z. Guo, T.-W. Huang, Y. Lin, GPU-accelerated static timing analysis, in *2020 IEEE/ACM International Conference On Computer Aided Design (ICCAD)* (2020), pp. 1–9
19. S. Bian, M. Shintani, M. Hiromoto, T. Sato, LSTA: learning-based static timing analysis for high-dimensional correlated on-chip variations, in *Proceedings of the 54th Annual Design Automation Conference 2017*, Series DAC '17 (Association for Computing Machinery, New York, NY, USA, 2017). Available https://doi.org/10.1145/3061639.3062280
20. A.B. Kahng, U. Mallappa, L. Saul, Using machine learning to predict path-based slack from graph-based timing analysis, in *2018 IEEE 36th International Conference on Computer Design (ICCD)* (2018), pp. 603–612
21. M.A. Savari, H. Jahanirad, NN-SSTA: a deep neural network approach for statistical static timing analysis, Expert Syst. Appl. **149**, 113309 (2020). Available https://www.sciencedirect.com/science/article/pii/S0957417420301342

Chapter 6
Area Analysis

Abstract At FPGA chip design stage, area analysis/estimation is essential, just like ASIC chips. State-of-the-art FPGA area estimation techniques will be discussed in this chapter. However, once the FPGA is manufactured, the chip area is fixed, and the "area problem" turns into resource utilization analysis, which can be accurately reported after implementation at FPGA application design stage.

6.1 Overview

After FPGA is manufactured, the chip area is fixed. Different application could occupy different resources. To predict how many resources an application could use under a given FPGA architecture is another topic [1–4]. Here we only discuss about FPGA area analysis/estimation at chip design stage. Its objective is to accurately predict the fabric area before the chip is manufactured. This process is important because it offers:

1. Cost estimation
   Estimating the area of a chip can provide an approximate idea of its manufacturing cost. The larger the chip, the more expensive it is to produce. Therefore, accurate area estimation helps in determining the cost of the chip, which can be useful in budgeting and decision-making.
2. Performance optimization
   The area of a chip can impact its performance. For instance, a smaller chip generally has a shorter propagation delay, which means it can operate faster. By estimating the area of the chip accurately, chip designers can optimize the layout of the chip to achieve the desired performance.
3. Yield prediction
   The yield of a chip refers to the number of good chips that can be obtained from a single wafer. Accurate area estimation helps in predicting the yield of a chip, which can be useful in planning the production process.

K. Tu et al., *FPGA EDA*, https://doi.org/10.1007/978-981-99-7755-0_6

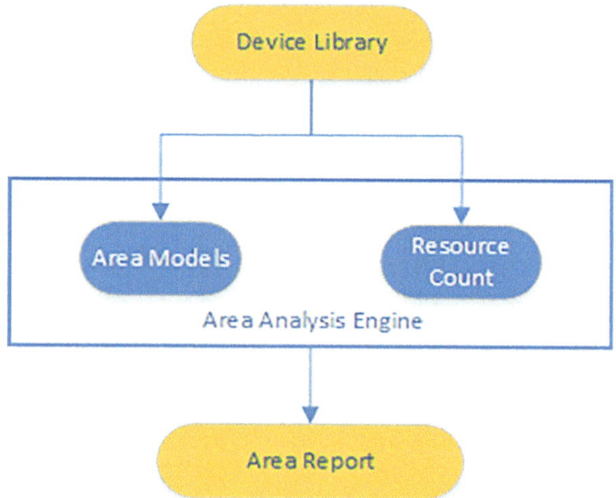

Fig. 6.1 Typical inputs and outputs in area analysis flow

A typical area analysis engine use device library (containing area models and device resource information) as its input and then output the chip area report (Fig. 6.1). Based on different area models, there can be different analysis methods: layout-based and minimum width transistor area (MWTA)-based.

6.2 Area Analysis Techniques

The most accurate way to analyze the area of an FPGA is layout-based technique—getting a complete layout. However, during the iterative FPGA design process, fast approximation techniques, for example, MWTA-based techniques, are also widely accepted to estimate the area of an FPGA.

1. Layout-based
   Once FPGA layout is complete, area of each part of the FPGA can be calculated (width times length). Example layouts are shown in (Fig. 6.2).
2. Minimum Width Transistor Area (MWTA)-based
   MWTA-based technique aims to fast approximate the FPGA area at transistor level and has been intensively studied in academia [6–10]. Based on MWTA model, which has been introduced in device (area) modeling chapter, this technique counts the transistors of the target circuit and compute the tile area by multiplying the minimum width transistor area count by the actual minimum width transistor area. Then, the final chip area is the product of the area of an individual tile and the number of tiles.
   The VTR model estimates the area of one minimum width transistor using the equation (Eq. 6.1).

$$Area(x) = 0.5 + 0.5x \tag{6.1}$$

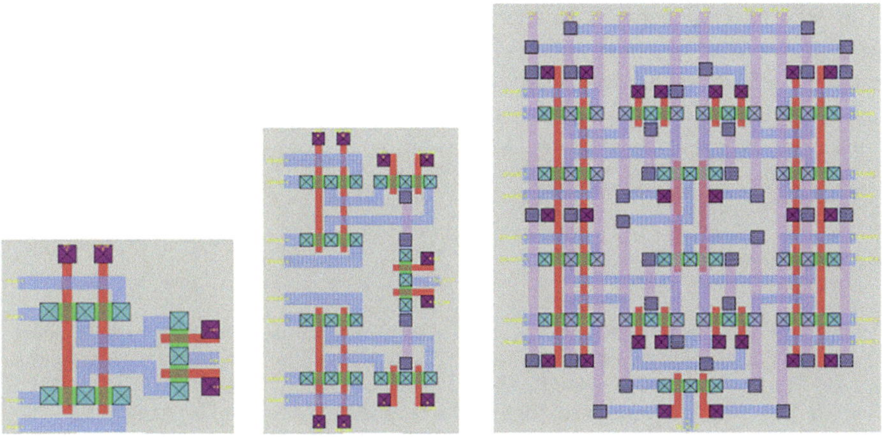

Fig. 6.2 Example layout of FPGA component: 2-LUT, 3-LUT, and 4-LUT [5]

where x is the transistor drive-strength. According to COFFE [11], this model over-predicts transistor area by up to 143% when compared to area measurements extracted from layouts with TSMC 65nm rules.

COFFE [10, 11] calculates the area of NMOS pass-transistors with (Eq. 6.2) and the area of CMOS transistors (e.g., inverters) with (Eq. 6.3).

$$\mathrm{Area}(x) = 0.447 + 0.128x + 0.391\sqrt{x} \tag{6.2}$$

$$\mathrm{Area}(x) = 0.518 + 0.127x + 0.428\sqrt{x} \tag{6.3}$$

However, works in [12] revised these models in a more realistic environment—taking metal layers into consideration. Correction factors are applied to adjust the equations above, and more accurate results are achieved in return.

6.3 Summary and Trends

Layout-based technique has set a golden benchmark for any other area estimation methods in terms of quality of results. However, accuracy and efficiency are trade off again in this scenario—more efficient yet less accurate methods are studied as a supplement to layout-based technique. Finding out chip area for each FPGA units and then sum them up is the easiest way to think of, no matter which type of area model is adopted.

MWTA-based techniques have set good examples for these efforts above. It dramatically increases the area analysis speed by using universal models, based on that, following studies try to find a better balance between accuracy and efficiency by adding more model details such as targeting a wider range of FPGA components and considering multiple metal layers [12, 13].

References

1. D. Kulkarni, W. Najjar, R. Rinker, F. Kurdahi, Fast area estimation to support compiler optimizations in FPGA-based reconfigurable systems, in *Proceedings in 10th Annual IEEE Symposium on Field-Programmable Custom Computing Machines* (2002), pp. 239–247
2. S.-K. Lam, W. Li, T. Srikanthan, High level area estimation of custom instructions for FPGA-based reconfigurable processors, in *2007 6th International Conference on Information, Communications & Signal Processing* (2007) pp. 1–5
3. M.B. Abdelhalim, S.E.-D. Habib, Fast FPGA-based delay estimation for a novel hardware/software partitioning scheme, in *2007 2nd International Design and Test Workshop* (2007), pp. 175–181
4. X. Wei, J. Chen, Q. Zhou, Y. Cai, J. Bian, X. Hong, Macromap: a technology mapping algorithm for heterogeneous FPGAs with effective area estimation, in *2008 International Conference on Field Programmable Logic and Applications* (2008), pp. 559–562
5. F.F. Khan, Towards accurate FPGA area models for FPGA architecture evaluation (2021). [Online]. Available: https://rshare.library.torontomu.ca/articles/thesis/Towards_accurate_FPGA_area_models_for_FPGA_architecture_evaluation/14660364
6. A. Marquardt, V. Betz, J. Rose, Speed and area tradeoffs in cluster-based FPGA architectures, *IEEE Transactions on Very Large Scale Integration Systems*, vol. 8, no. 1 (2000), p. 84–93 [Online]. Available: https://doi.org/10.1109/92.820764
7. E. Ahmed, J. Rose, The effect of LUT and cluster size on deep-submicron FPGA performance and density, *IEEE Transactions on Very Large Scale Integration (VLSI) Systems*, vol. 12, no. 3 (2004), pp. 288–298
8. A.M. Smith, G.A. Constantinides, P.Y.K. Cheung, Area estimation and optimisation of FPGA routing fabrics, in *2009 International Conference on Field Programmable Logic and Applications* (2009), pp. 256–261
9. I. Kuon, J. Rose, Exploring area and delay tradeoffs in FPGAs with architecture and automated transistor design, *IEEE Transactions on Very Large Scale Integration (VLSI) Systems*, vol. 19, no. 1 (2011), pp. 71–84
10. S. Yazdanshenas, V. Betz, Coffe 2: automatic modelling and optimization of complex and heterogeneous FPGA architectures, vol. 12, no. 1 (2019). [Online]. Available: https://doi.org/10.1145/3301298
11. C. Chiasson, V. Betz, Coffe: fully-automated transistor sizing for FPGAs, in *2013 International Conference on Field-Programmable Technology (FPT)* (2013), pp. 34–41
12. M. Al-Qawasmi, A.G. Ye, An investigation of the accuracy of the vpr and coffe area models in predicting the layout area of FPGA lookup tables, in *2020 SoutheastCon* (2020), pp. 1–9
13. A.M. Smith, G.A. Constantinides, P.Y.K. Cheung, Area estimation and optimisation of FPGA routing fabrics, in *2009 International Conference on Field Programmable Logic and Applications* (2009), pp. 256–261

Part IV
FPGA Chip Design EDA

Chapter 7
Semi-custom EDA

Abstract In modern computing systems, FPGAs are used as dedicated programmable accelerators (Che et al. [1], Zhang et al. [2], Cong et al. [3]). General-purpose FPGAs are well optimized to fit a wide range of applications with a reasonable trade-off on performance, power, and area, but are seriously sub-optimal in application-specific contexts (Cong et al. [3], Neshatpour et al. [4]). In such case, customized FPGA architectures, which are highly tailored for a specific set of applications as well as seamless integration to other computing resources in the system, become a proper solution. However, developing a FPGA layout through full custom approaches is a time-consuming process even for industrial vendors, whose may take years to finalize (Greenhill et al. [5]). In addition, design tools such as mapping algorithms and bitstream generation have to be customized for different FPGA architectures, which lead to another time-consuming development task. Driven by the strong need, fast prototyping technology for customize FPGAs, especially semi-custom design approaches, has been insensitively researched in recent years. As such, development cycles of custom FPGAs can be comparable to modern ASICs, which opens the door to tightly integrating FPGAs to SoCs. In this section, we will first review existing EDA tools and then focus on critical EDA techniques that enable semi-custom designed FPGAs.

7.1 Overview

In the past two decades, fast prototyping techniques for customized FPGA architectures have been proven by many researches through semi-custom design flows [6–14]. These works share the same principles when generating FPGA layouts:

K. Tu et al., *FPGA EDA*, https://doi.org/10.1007/978-981-99-7755-0_7

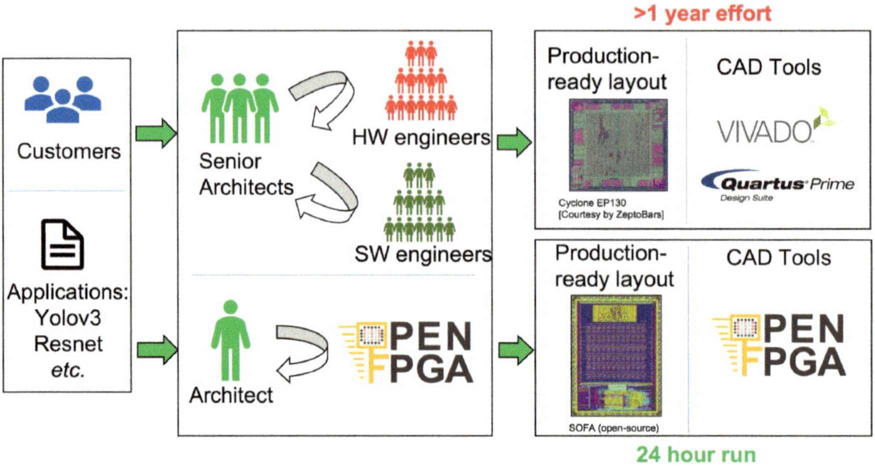

Fig. 7.1 An illustrative example that compares on engineering time and effort to prototype an FPGA using OpenFPGA (an open-source EDA tool that enables semi-custom approaches) and full-custom approaches

1. Model an FPGA architecture in synthesizable HDL netlists.
2. Use sophisticated ASIC design tools to implement the HDL netlists into physical layouts.

As illustrated in Fig. 7.1, the fast prototyping technology through semi-custom design flows accelerates and automates the development process of FPGAs.

Early works rely on handcrafted HDL netlists for FPGA architectures which even include low-level details down to transistor-level circuit designs [6, 7]. However, such methodology requires still significant manual effort, being inefficient in designing diverse FPGA fabrics targeting domain-specific applications. Moreover, early works focus only on developing fabric generators without associated compiler support, e.g., HDL-to-Bitstream generation [6, 7]. Recent works aim to build "FPGA generators" in the similar concept as the memory compilers in ASIC world [8–14]. The FPGA generators integrate both netlist generators and bitstream generators in a unified framework, on top of the well-known FPGA architecture exploration tool, e.g., VTR [15, 16]. Major technical features of existing FPGA generators are summarized in Table 7.1.

However, to implement production quality FPGA fabrics, layout generation is only a small part (Fig. 7.2), when compared to other essential aspects, such as testbench generator and bitstream support. For example, to verify the correctness of FPGA fabrics before taping out, design verification is a mandatory step. Note that design verification for FPGAs is mainly a software problem rather than a hardware problem, as functionality of an FPGA is determined by a bitstream file. Therefore, to ensure a high coverage in verification, a number of bitstream files are required to verify different operating modes and utilization rates of an FPGA device. As a result, a

Table 7.1 Comparison on EDA tools enabling semi-custom FPGA design

Tool/metric	Open source	Architecture language	Netlist generation	Bitstream generation	Testbench generation	SDC generation
Kuon et al. [6]	×	✓	Automatic[a]	×	×	×
Ova et al. [7]	×	×	Hand-crafted	×	×	×
Archipelago [10]	✓	×	Automatic	✓	×	×
Anderson et al. [8, 9]	×	✓	Automatic	✓	✓	✓
Mohan et al. [13]	×	✓	Automatic	✓	✓	✓
PRGA [11]	✓	✓	Automatic	✓	×	×
FABulous [12]	✓	✓	Automatic[b]	✓	✓	×
OpenFPGA [14]	✓	✓	Automatic	✓	✓	✓

[a]Only netlists of a tile is automatically generated
[b]Netlists of primitive circuits, e.g., LUT and routing multiplexers, have to be hand crafted

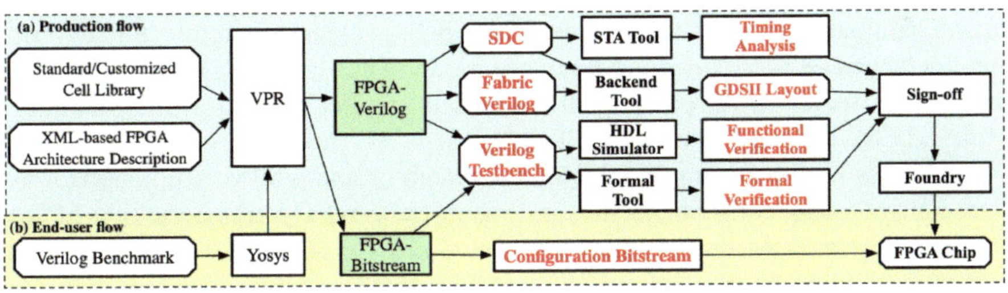

Fig. 7.2 Semi-custom design flow for FPGA fabrics: **a** production flow and **b** end-user flow

functional HDL-to-Bitstream generator is a required component, being as important as a netlist generator. In addition, a testbench generator is required to simulate the bitstream downloading w.r.t. a configuration circuits, as well as check the functional correctness of an FPGA under different I/O mapping and bitstreams. Actually, the complexity of a HDL-to-Bitstream flow is significantly higher than a netlist generator, which covers many NP hard problems in EDA, such as placement and routing. In recent years, with the growth of open-source HDL-to-Bitstream tools, design verification has been seriously considered and included in recent EDA tools, as shown in Table 7.1. In short, design verification for FPGA should not only validate the correctness of layout but also the correctness of associated software tool chains.

Beyond the essential components, to enable high-quality FPGA fabrics, timing constraints for physical design are critical. Nowadays, timing constraints are typically in the *Synopsys Design Constraints*(SDC) format, which are used to constrain timing paths when ASIC tools generate FPGA layouts. Without timing constraints, pin-to-pin delays, such as LUT delays and routing delay, may be too large to satisfy the target performance of an FPGA. Note that, a key difference between FPGAs and ASICs on timing paths is that an FPGA only has critical paths when mapped to a specific HDL design. When implementing FPGA layouts, timing constraints cannot be biased to an HDL design because it may probably cause performance degradation on another HDL design. Therefore, the principle of the timing constraints is keep pin-to-pin delays on each timing path as uniform as possible, which indicates that every timing path is critical. Considering the large number of timing paths in a FPGA fabric, a SDC generator is required to avoid huge manual effort. Nowadays, to achieve high-performance FPGA fabrics, SDC generators are available in semi-custom EDA tool chains (Table 7.1).

As architecture of FPGAs can be really different depending on their application context, a key value of FPGA generators is to support versatile FPGA architectures. Therefore, FPGA architecture description languages are needed to model complicated and large-scale FPGA device in compact and human readable representations. By leveraging the *University of Toronto FPGA Architecture Language*(UTFAL) [17], FPGA generators can convert a high-level FPGA description to synthesizable HDL netlists, and then implement layouts through ASIC design tools. Thanks to UTFAL's enriched syntax, FPGA generators can support a wide range of FPGA architectures. To unlock more possibility in device modeling, extended architecture description language (set of architecture guidance models) has been proposed [18], In this chapter, we focus on introducing the extended architecture description language while UTFAL has been covered in Chap. 2.

In short, a netlist generator, a bitstream generator and a testbench generator are three indispensable components in a basic semi-custom EDA framework for FPGA, with which designers can accomplish a functional FPGA fabric. However, as the growing needs of domain-specific FPGA fabrics, an expressive architecture language is now becoming important, because it is a must-have for designers to rapidly evaluate and prototype innovative FPGA architectures. As researchers have proven the feasibility of FPGA generators with silicon results (Fig. 7.3), future trends lie on improving PPA of the FPGA fabrics. This drives SDC generators to be an strategically important tool, which can constrain PPA of each segment in an FPGA fabric through semi-custom design tools w.r.t. performance goals.

7.2 Extended Architecture Description Language

In this part, we focus on the extended architecture description language(set of architecture guidance models conceptualized in Chap. 2) adopted by the OpenFPGA framework [18]. Other architecture description languages may have different syntax when modeling FPGA fabrics but share similar principles [11, 12]. Therefore, we focus more on general principles when designing an FPGA architecture description language than detailed syntax, with which we believe it is easier for readers to understand other architecture description languages.

UTFAL is designed for a detailed logical representation of FPGA architectures, providing sufficient information for EDA engines to perform packing, placement, and routing. However, to enable netlist generation and bitstream generation, a detailed

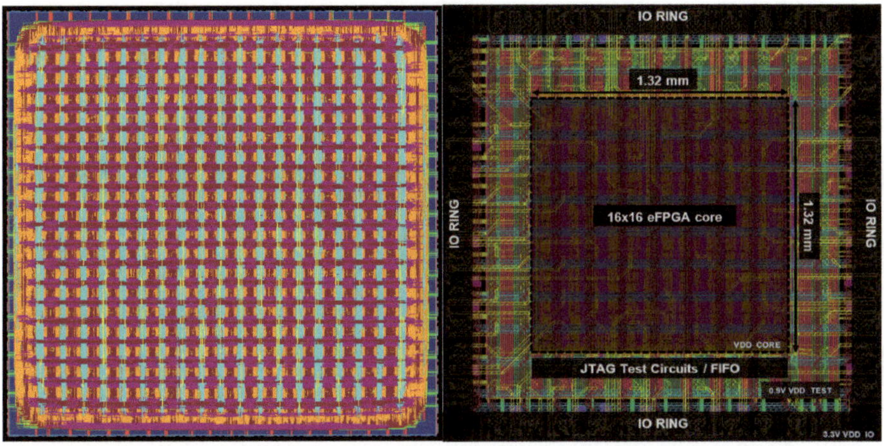

(a) A 20x20 FPGA fabric (courtesy by [8]) (b) A 16x16 FPGA chip (courtesy by [13])

Fig. 7.3 Showcase FPGA layouts through semi-custom design approaches

physical representation of complete FPGA fabric is required. The extended architecture description language is designed to provide supplementary information on top of the UTFAL. It fills the blank of UTFAL when modeling circuit-level implementation of programmable resources (see Sect. 7.2.1), physical mode of programmable blocks (see Sect. 7.2.2), and configuration scheme (see Sect. 7.2.2). Therefore, the extended architecture description language is complementary to UTFAL without overlapping in syntax and information. Similar to UTFAL, the extended architecture description language is XML-based. Full documentation about UTFAL and the extended architecture description language is available on [19, 20], respectively.

7.2.1 Circuit Modeling

As circuit design is a dominant factor impacting FPGA's PPA, the extended architecture description language provides enriched syntax to model circuit-level details of primitives in FPGAs, e.g., LUT, routing multiplexers. Figure 7.2 illustrates the different focus on modeling LUTs and routing multiplexers between UTFAL and the extended architecture description language. For EDA usage only, primitives can be treated as a black box with limited information, e.g., number of ports, port direction as well as pin-to-pin delays. However, to generate netlists, detailed circuit designs of primitives have to be modeled. On the other side, upon practical applications, hardware engineers may select various circuits to implement their FPGA fabrics. For instance, a ultra-low-power FPGA may be built with ultra-low-power circuit cells while a high-performance FPGA may use absolutely different circuit cells. As a result, the extended architecture description language is capable of modeling highly

```
<!-- Fracturable LUT modeled in XML -->
<circuit_model type="lut" name="frac_lut4" verilog_netlist="lut.v">
  <design_technology fracturable_lut="true"/>
  <pass_gate_logic circuit_model_name="tgate"/>
  <port type="input" prefix="lut_i" size="4" tri_state_map="---1"/>
  <port type="output" prefix="lut3_o" size="2" lut_frac_level="3"
lut_output_mask="0,1"/>
  <port type="output" prefix="lut4_o" size="1"
lut_output_mask="0"/>
  <port type="sram" prefix="sram" size="16"/>
  <port type="sram" prefix="mode" size="1" mode_select="true"/>
</circuit_model>
```

Fig. 7.4 Examples of extended XML syntax for LUTs

flexible circuit design topology even down to transistor level and allows designers to customize any component in an FPGA.

Among the programmable resources in an FPGA, there are two types of circuits whose structures have prominently impact on PPA and bitstream generator: LUTs and routing multiplexers. LUTs are used to implement logic functions while routing multiplexers are used to route signals between LUTs. In some FPGA devices, LUTs and routing multiplexers take 90% of chip area, critical path delays, and power consumption [21]. The choice of the circuit implementation may also impact the PPA of standalone circuit by 2× [22]. Therefore, the extended architecture description language provides fruitful syntax to support diverse circuit design topology and details for LUTs and routing multiplexers.

Table 7.3 lists the mainstream circuit topology for LUTs and routing multiplexers that are frequently used by modern FPGAs. Figure 7.4 shows an example about how the extended architecture description language models the internal structure of a fracturable 4-input LUT. Users can specify which inputs are disabled during fracturable mode in the XML property `tri_state_map`. The levels and positions of fracturable outputs can be freely defined through the XML properties `lut_frac_level` and `lut_output_mask`. To support mode switching of fracturable LUTs, the port map includes a special port `mode` rather than the regular configuration port. Figure 7.5 shows another example about how a tree-like 4-input routing multiplexer (see Table 7.2 for schematic) is modeled by the extended architecture description language. The multiplexing structures can be customized through an XML property `structure`. Note that both input, output and even intermediate buffers can be customized through XML syntax, which are needed for LUTs and routing multiplexers in different location of an FPGA. With these modeling, a netlist generator can output RTL and even gate-level netlists for the LUTs and routing multiplexers, meanwhile bitstream generator can decode configuration bits.

In addition to the detailed modeling, black-box modeling is also supported, where users can provide their own circuit implementation for primitives. When black-box modeling is adopt, the path to netlist should be defined through the XML property `verilog_netlist`, and only necessary information such as port list is required.

```
<!-- Tree-like MUX modeled in XML -->
<circuit_model type="mux" name="mux_tree"/>
  <design_technology structure="tree"/>
  <pass_gate_logic circuit_model_name="STD_CELL_MUX2"/>
  <port type="input" prefix="mux_in" size="4"/>
  <port type="output" prefix="mux_out" size="1"/>
  <port type="sram" prefix="mux_sram" size="2"/>
</circuit_model>
```

Fig. 7.5 Examples of extended XML syntax for MUXes

Such modeling is also frequently used as modern FPGAs are built with various third-party IPs, e.g., *Digital Signal Processor* (DSP), *Random Access Memory* (RAM) and *Serializer/Deserializer* (SerDes).

7.2.2 Physical Mode Modeling

To simplify EDA algorithms, UTFAL focus on compact description of *Logic Element* (LE) architectures instead of a complete schematic-level representation. For instance, a complex multi-mode LE in Fig. 7.6a is modeled by multiple abstract-level operating modes in Fig. 7.6b, c. The abstraction indeed eases the EDA algorithms in mapping to FPGA resources but hides important details required by netlist and bitstream generation for the physical LEs. For example, netlist generators cannot identify which mode in Fig. 7.6 denotes the physical implementation of the LE. Bitstream generators may miss configuration bits to be decoded in physical mode when the operating modes in Fig. 7.6b, c only include a part of programmable routing resources. Moreover, configuration bits of an operating mode should be properly reorganized for the physical mode. For example, the configuration bits of the two `3-LUT` in Fig. 7.6c should be mapped to the fracturable `4-LUT` in Fig. 7.6a. Without a detailed circuit-level implementation of the fracturable `4-LUT`, bitstream generators cannot even decode configuration bits of the two `3-LUT` from logic synthesis results.

Therefore, to enable both netlist and bitstream generators, extended syntax is developed to

1. distinguish between physical mode and operating modes;
2. link the components in the various operating modes to physical mode
3. establish the relationship between primitives in physical mode and their circuit-level modeling (see Sect. 7.2.1).

To be intuitive, we take the example of the multi-mode CLB shown in Fig. 7.6 and present XML description in Fig. 7.7. The physical implementation of the LE is specified to be the mode `phy`, through syntax `physical_mode_name`. The detailed architecture of the physical LE follows the same style as the UTFAL. Under the physical mode, users can link primitive blocks to circuit implementations using a XML property `circuit_model_name`. Figure 7.7 shows how a

Table 7.2 Different objectives between UTFAL and extended architecture description language: logical vs. physical modeling

UTFAL	Extended architecture description language
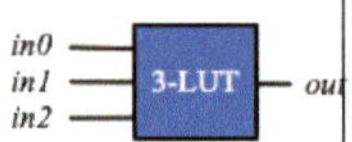	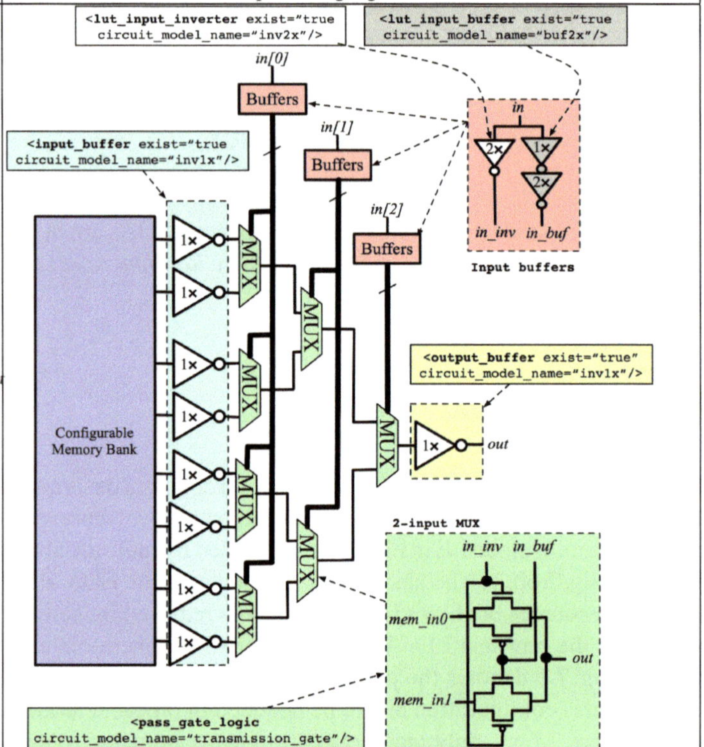
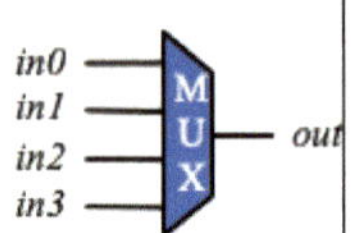	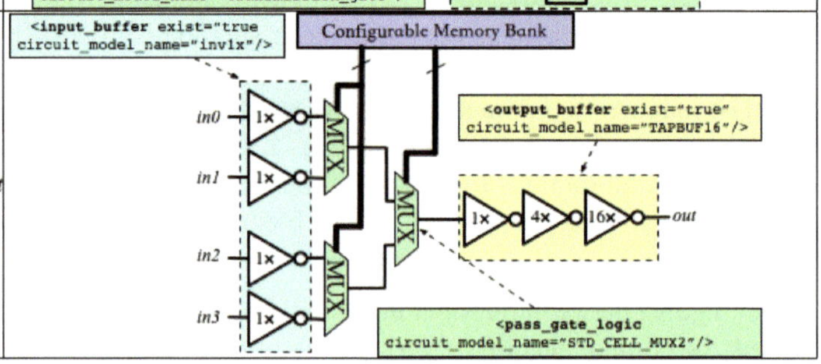

Table 7.3 Various circuit designs of LUTs and routing multiplexers

Circuit	Design topology
LUT	1. Single-output LUTs
	2. Fracturable (multi-output) LUTs
	3. LUT with hard logic, e.g., carry
	4. LUT built with standard cells
	5. LUT with RAM/ROM
Routing multiplexer	1. One-level multiplexer
	2. Multi-level multiplexer
	3. Tree-like multiplexer
	4. Standard-cell multiplexer
	5. Multiplexer with local encoder
	6. Multiplexer with constant input

*Input and output buffering can be fully customized for both circuits

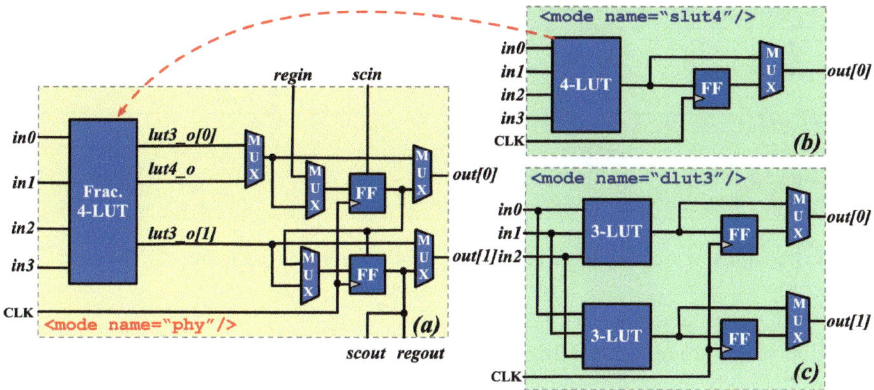

Fig. 7.6 **a** Physical implementation of a LE and **b**, **c** two operating modes

fracturable LUT `flut` is linked to a defined circuit model `frac_lut4` in Fig. 7.4. Under the operating modes, each virtual `pb_type` has to be linked to its physical implementation through XML properties `physical_pb_type_name` and `physical_mode_port`. Consider the example in Fig. 7.7, the operating modes `dlut3` and `slut4`, which correspond to the illustration in Fig. 7.6b, c, are linked to the physical mode `phy` which correspond to the illustration in Fig. 7.6a. The inputs `in` and outputs `out` of the pb_type `lut4` in mode `slut4` are linked to the inputs `in[0:3]` and outputs `lut4_o` of the pb_type `flut` in its physical mode `phy`, as highlighted by red dash lines in Fig. 7.6. XML syntax `mode_bits` allows users to customize the configuration bits applied to fracturable LUTs in any operating mode. For example, in Fig. 7.7, when the `lut4` is used, the `mode_bits="1"` will be applied to the port `mode` of its physical module `frac_lut4` in Fig. 7.4. As such, without modifying packing or synthesis engines, the XML syntax can map the con-

```
<!-- physical pb_type binding in logic element -->
<!-- Specify the physical mode -->
<pb_type name="le" physical_mode_name="phy"/>
<!-- Specify the circuit model for the primitives in physical
mode -->
<pb_type name="le[phy].flut" circuit_model_name="frac_lut4"/>
<pb_type name="le[phy].ff" circuit_model_name="sdff"/>
<!-- Bind operating modes to physical modes -->
<pb_type name="le[dlut3].lut3" physical_pb_type_name="le[phy].flut"
mode_bits="0">
  <port name="in" physical_mode_port="in[0:2]"/>
  <port name="out" physical_mode_port="lut3_o"/>
</pb_type>
<pb_type name="le[dlut3].ff" physical_pb_type_name="le[phy].ff"/>
<pb_type name="le[slut4].lut4" physical_pb_type_name="le[phy].flut"
mode_bits="1">
  <port name="in" physical_mode_port="in[0:3]"/>
  <port name="out" physical_mode_port="lut4_o"/>
</pb_type>
<pb_type name="le[slut4].ff" physical_pb_type_name="le[phy].ff"/>
```

Fig. 7.7 Examples of extended XML syntax for a LE

```
<configuration_protocol>
  <organization type="memory_bank" circuit_model_name="sram"
num_regions="4"/>
</configuration_protocol>
```

Fig. 7.8 Examples of memory-bank-based configuration protocol modeling

figuration bits from any operating mode to its physical implementation. In addition, such multi-mode modeling enable users to define a simplified BLE architecture in operating modes than physical mode, which reduces CPU time for packing.

7.2.3 Configuration Protocol

Programmable resources in an FPGA have to be configured through a protocol. However, configuration protocols are not modeled in UTFAL because they are well decoupled from packing, placement, and routing algorithms. Configuration scheme directly impacts bitstream generators, which is essential to a complete tool chain. More importantly, configuration protocol could be really different in FPGAs, depending on the application context. Extended architecture description language is developed to support versatile configuration protocols. Figure 7.8 shows an example of modeling a memory-bank-based configuration protocol, where other types of configuration protocol can be specified through XML property `type`. Through memory banks, each configuration memory cell can be accessed by enabling dedicated *Bit-Line* (BL) and *Word-Line* (WL). Note that the circuit implementation of a memory cell can be not

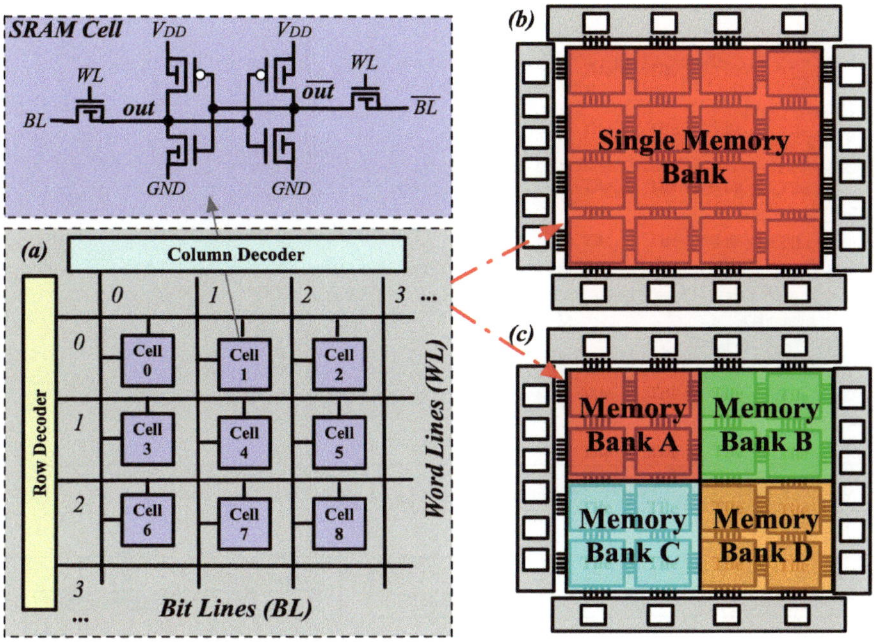

Fig. 7.9 Example of **a**, a memory organization using decoders; **b** single memory bank across the fabric; and **c** multiple memory banks across the fabric

limited to a SRAM, as shown in Fig. 7.9. For example, flip-flops or latches can also used as the fundamental cell in memory banks. The circuit model of configuration memory cell can be specified through XML property `circuit_model_name`. In addition, as FPGA size grows, multiple configuration regions are adapted to avoid long configuration time as well as challenges in physical design due to large parasitic in BL/WL interconnection. Figure 7.9b, c shows illustrative examples of single-region and 4-region memory banks, respectively. Therefore, the number of configuration regions can be customized through the XML property `num_regions`. Note that other configuration protocols, such as configuration chains and frame-based, are parameterized as memory banks, where different number of regions and various circuit implementation may also be applied.

In practice, configuration scheme for each tile or lower level primitive may need full customization. Take the example of memory bank, chip designer may need to customize which tiles to share BLs and WLs, in order to optimize in physical design and configuration time. Figure 7.10 shows an example file where designers can specify BL and WL sharing for each tile in each configuration region of an FPGA fabric. Two tiles share the same BL when their column index are same. Two tiles share the same WL when their row index are same. Consider the example in Fig. 7.10, the two tiles `grid_io_bottom_1__0_` and `grid_io_bottom_2__0_` are configured by the same WL but through two different BLs, where the BLs and WLs

```
<fabric_key>
  <region id="0">
    <<key id="0" alias="grid_io_bottom_1__0_" column="0" row="0"/>
    <<key id="0" alias="grid_io_bottom_2__0_" column="1" row="0"/>
  </region>
  <region id="1">
    <<key id="0" alias="grid_io_right_3__1_" column="2" row="1"/>
    <<key id="0" alias="grid_io_right_3__2_" column="2" row="2"/>
  </region>
  <region id="2">
    <<key id="0" alias="grid_clb_1__1_" column="0" row="1"/>
    <<key id="0" alias="grid_clb_1__2_" column="0" row="2"/>
  </region>
  <region id="3">
    <<key id="0" alias="grid_clb_2__1_" column="1" row="1"/>
    <<key id="0" alias="grid_clb_2__2_" column="1" row="2"/>
  </region>
</fabric_key>
```

Fig. 7.10 Examples of fabric key file modeling BL/WL sharing

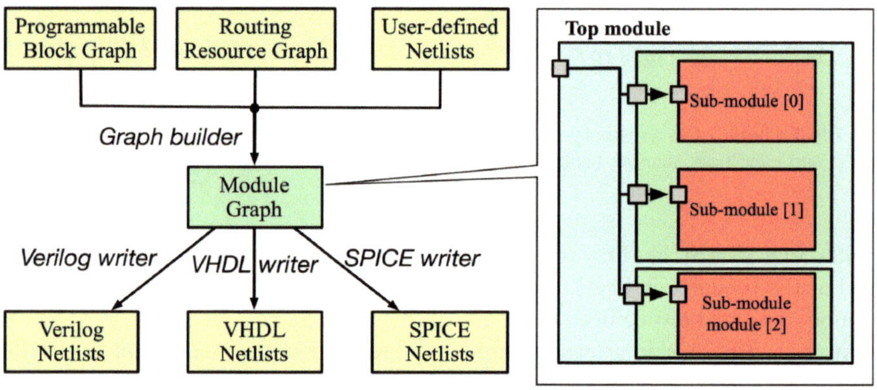

Fig. 7.11 Flowchart of netlist generator and graph-based modeling for modules

are controlled by `region 0`. For each region, different set of BLs and WLs are used to control the tiles under it. A tile can only be controlled by a configuration region. We refer interested reader to [20] for details.

7.3 Netlist Generator

As a cornerstone of the semi-custom design tools, netlist generators aim to translate a high-level architecture description to HDL netlists which can be adapted by ASIC tools to implement physical layouts. In early works, netlist generators is a simple

HDL code generator [6–10], which outputs internal device modeling to a synthesizable HDL format in a straightforward way. However, such native HDL translation of FPGA fabric imposes strong limitation when implementing physical layouts. For example, considering the HDL netlist which model a complete routing fabric as a flatten graph, the file sizes of netlists increase exponentially when FPGA size increases, which causes a long runtime in physical design. Furthermore, flatten netlists force a high design complexity when implementing an FPGA fabric, since a 4K-LUT FPGA may contain 8+ millions of logic gates. As a result, the physical design runtime of a medium sized FPGA is more than 24 h [8], while the physical design may fail for large sized FPGAs [23]. Modern netlist generators are designed to not only a simple code generator but also contain many features which make outputted netlists to be:

1. physical design friendly;
2. compatible with multiple HDL format and their standards;
3. human-readable, easy to debug and backtrace errors.

To enable these features, as depicted in Fig. 7.11, the implementation of the netlist generators is based on two steps:

1. Create a graph of modules which represent the complete FPGA fabric;
2. Build a number of netlist writers which output the module graph into selected file formats.

In the graph-based modeling, the whole FPGA fabric is represented as a tree of modules and their instances, as shown in Fig. 7.11. Modeling an FPGA fabric in a graph allows EDA tool to easily adjust hierarchy of netlists. For example, through graph merging, sub-modules can be merged which unlocks more opportunity in physical design optimization. It is also straightforward to profile the FPGA fabric, e.g., get the depth of netlists, count number of unique modules, etc., which can provide critical information for physical designers. A graph can be outputted to different file formats through various netlist writers, such as Verilog writer. As such, netlist writers consider a graph as an input, being decoupled from rest of engines in netlist generators. This can avoid massive code changes in core engine when developing a new netlist writer.

The auto-generated fabric netlists include both a programmable fabric with configuration protocol embedded. To be physical design friendly, netlist generators are capable of outputting netlist in different levels, e.g., *Register-Transfer Level* (RTL) and *Gate-level* (GL). Netlists at different levels of details unlock optimization opportunities through different design flows. As illustrated in Fig. 7.12, RTL (behavioral) netlists can be optimized through synthesis tools to standard cells and then physically implemented to layouts. Alternatively, GL (denoted as technology-mapped in Fig. 7.12) netlists are preferred as an direct input to physical design tool, when chip designers require specific standard cells to implement primitive circuits which are not synthesizable. The choice of design flows really depends on the PPA requirements and expertise of chip designers. For example, for ultra-high-performance FPGA, some specific cells are required in gate-level netlists and synthesis should be skipped.

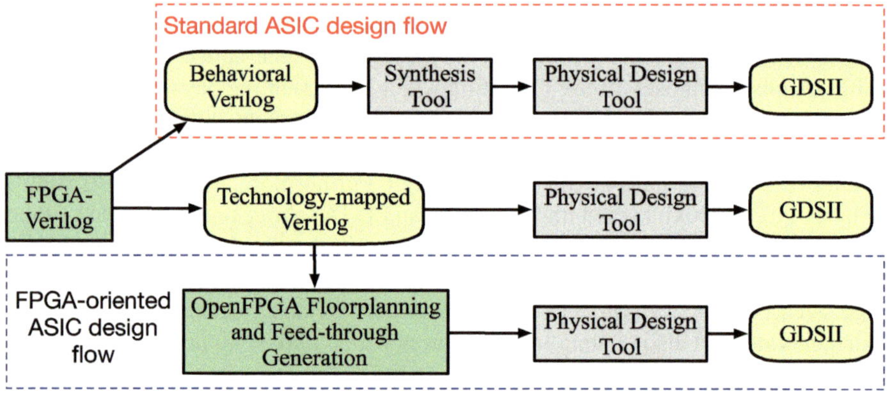

Fig. 7.12 An example of physical-design-friendly netlist generators

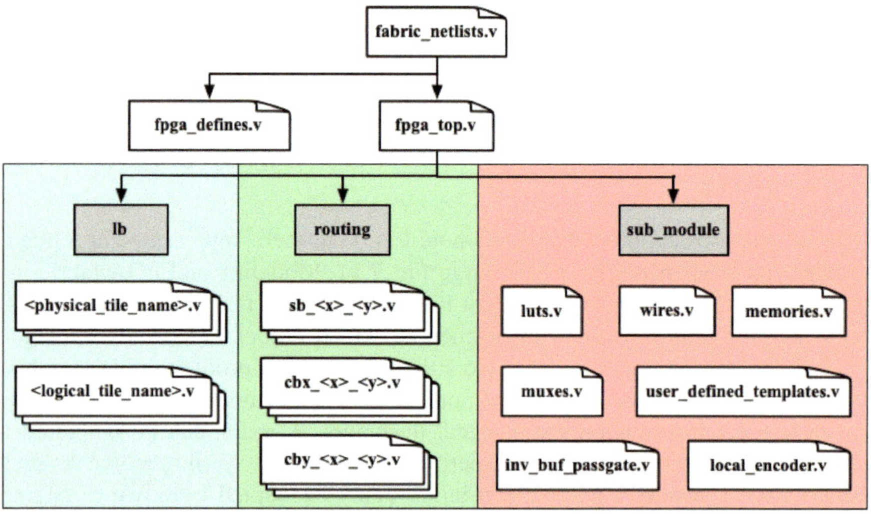

Fig. 7.13 An example of hierarchical Verilog netlists modeling a FPGA fabric

On the other side, the hierarchy of netlists also impact the physical design significantly. Figure 7.13 illustrates an example of Verilog netlists which are outputted by the OpenFPGA, which models a complete FPGA fabric in a hierarchical way. Note that highly hierarchical fabrics are generated, where large FPGAs can be built with a small number of repeatable tiles including routing blocks. Tiles and routing blocks are built with common primitive blocks, located in the `sub_module` directory, which can maximize the reuse of primitive netlists. Repeatable tiles can efficiently reduce the file sizes, total runtime, and design complexity of physical design flow. For example, in a physical design methodology, only unique tiles are placed and

```
<pb_type name="clb">
  <input name="I" num_pins="10" equivalent="full"/>
  <output name="O" num_pins="4" equivalent="none"/>
  <clock name="clk" num_pins="1"/>
<!-- child pb_type may be defined underneath -->
<pb_type>
```

(a) An examples of a CLB definition in UTFAL

```
module logical_tile_clb_mode_clb_(prog_clk, clb_I, clb_clk,
ccff_head, clb_O, ccff_tail);
  input [0:0] prog_clk;
  input [0:9] clb_I;
  input [0:0] clb_clk;
  input [0:0] ccff_head;
  output [0:3] clb_O;
  output [0:0] ccff_tail;
<!-- internal structure of clb is omitted -->
endmodule
```

(b) An example of a CLB in auto-generated Verilog netlists

Fig. 7.14 An example of auto-generated human-readable netlists corresponding to architecture definition

routed, while the top-level fabric is only an assemble of the tiles which are treated as black boxes [23].

Note that different physical design tools may require different HDL formats and their specific variants. Verilog is a popular HDL format for most physical design tools, while VHDL is more popular as a strict behavioral modeling for FPGA fabrics. Modern netlist generators include various netlist writers to convert a graph representation of FPGA fabric to the file format which meets downstream tool requirements. Even when considering Verilog format, various netlist styles may be demanded, in order to be compatible with latest Verilog standards. For instance, the syntax `default_nettype` is introduced to force strict wire definition in Verilog 2001. Supporting diverse syntax allows the auto-generated netlists to be more human readable and easier to back-trace errors for chip designers, especially when there are implementation errors during physical design flow. To further improve readability of outputted netlists, names of modules, ports, and nets should be human readable and correspond to architecture description. Figure 7.14 shows an example how the outputted netlist can be easy to correlated to the architecture description. In Fig. 7.14a, a programmable block `clb` with two input ports and one output is defined using the UTFAL. Figure 7.14b presents the Verilog codes which are outputted by OpenFPGA, corresponding to the programmable block `clb`. The port name and port size are consistent between the architecture description and the netlists, through which chip designer can backtrace the changes in netlists to a specific portion of architecture file. For instance, the port `I` of `clb` in Fig. 7.14a is named as `clb_I` in Fig. 7.14b.

We refer interested readers to [20] for a detailed implementation of netlist generator.

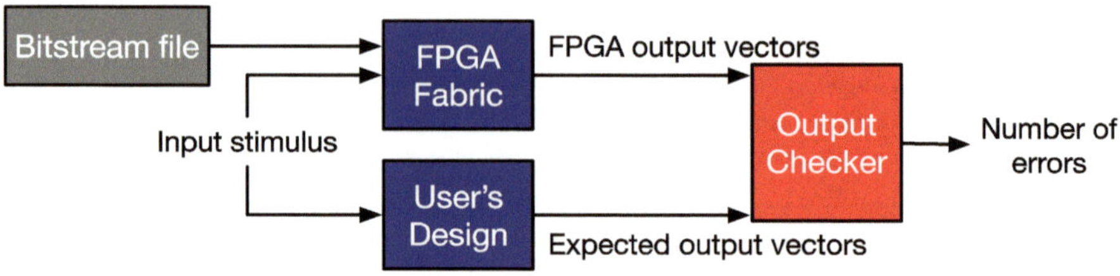

Fig. 7.15 Principles of Verilog testbenches: (1) using common input stimuli; (2) applying bitstream; (3) checking output vectors

Table 7.4 Auto-generated testbench features

Testbench	Runtime	Test vector	Test coverage
Full	Long	Random stimulus	Full fabric
Preconfigured	Short	Random stimulus/formal method	Programmable fabric only

7.4 Testbench Generator

It is essential to validate the correctness of FPGA fabrics before tape-out. However, a key difference between the design verification for FPGAs and ASICs lies on bitstreams. As highlighted in Fig. 7.15, an FPGA carries a specific functionality only when an associated bitstream is loaded. To ensure a high verification coverage, chip designers need a number of bitstream files, each of which is designed to validate a specific part of the FPGA. The bitstream files could be either synthetic (not synthesizable through HDL-to-bitstream tools) or based on a user's RTL design. To validate the various bitstream on an FPGA, testbenches have to be generated with dedicated I/O mapping for each configuration. Note that for most applications, only part of FPGA I/Os are used and for each application, each FPGA I/O may be used in a different way. Testbench generators assign the I/O mapping based on the results from HDL-to-Bitstream results. To enable self-testing, the FPGA and user's RTL design (simulated using an HDL simulator) are driven by the same input stimuli, and any mismatch on their outputs are reported as errors.

To trade-off runtime and coverage, as listed in Table 7.4, two types of testbenches are typically generated to validate the correctness of the fabric before tape-out: full and preconfigured. Full testbench aims at simulating an entire FPGA operating period, consisting of two phases:

1. the configuration phase, where the bitstream file is loaded to the programmable fabric through a configuration protocol, as highlighted by the green rectangle of Fig. 7.16;
2. the operating phase, where random input vectors are applied to drive both *Devices Under Test* (DUTs), as highlighted by the red rectangle of Fig. 7.16.

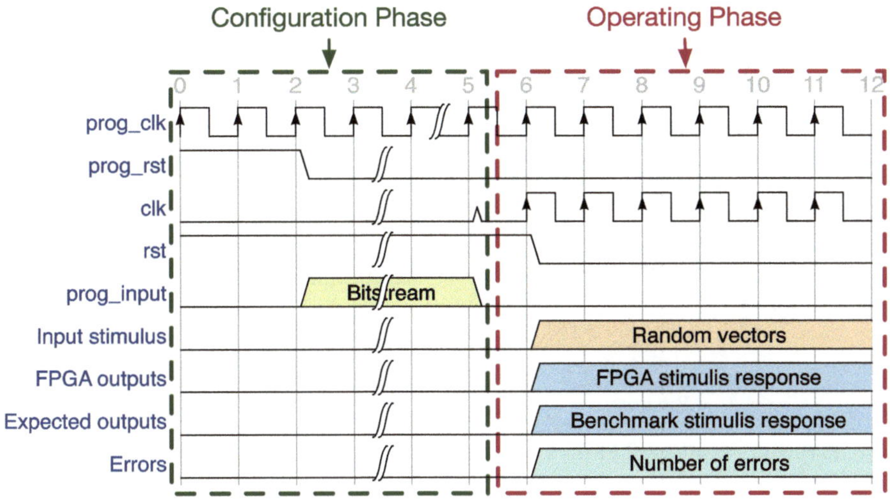

Fig. 7.16 Illustration on the waveforms in full testbench

Using the full testbench, chip designers can validate both the configuration circuits and programming fabric of an FPGA. However, the random testing vectors used in the full testbench may result in only a small set of functional coverage. On the other side, as the bitstream size increases exponentially with the FPGA size, the number of clock cycles required to load the bitstream becomes a dominating factor (more than 90%) in the verification runtime. For instance, HDL simulation of a full testbench including a 800k-bit bitstream consumes a 24-hour runtime when using a commercial state-of-the-art simulator. In short, even there are significant limitations, the full testbench remains a must-run verification, since it fully validates the configuration protocol.

To improve the coverage, the preconfigured testbench is proposed, which skips the time-consuming configuration phase and focus on the operating phase. As a result, sufficient number of testing vectors can be applied to ensure functional correctness of a mapped FPGA design, while simulation runtime is fairly small. To apply testing vectors to mapped I/Os of an FPGA, a preconfigured FPGA, which is instantiated with the user's bitstream, is encapsulated with the same port mapping as the user's RTL design, as illustrated in Fig. 7.17. Note that beyond the functional verification show in Fig. 7.15, the preconfigured FPGA module can be also fed to a formal tool for a 100% coverage formal verification against user's RTL design. Compared to the full testbench, the preconfigured testbench significantly accelerates the functional verification especially for large FPGAs.

We believe that with proper use of the two types of testbenches, the verification process for FPGAs can be significantly simplified or even automated.

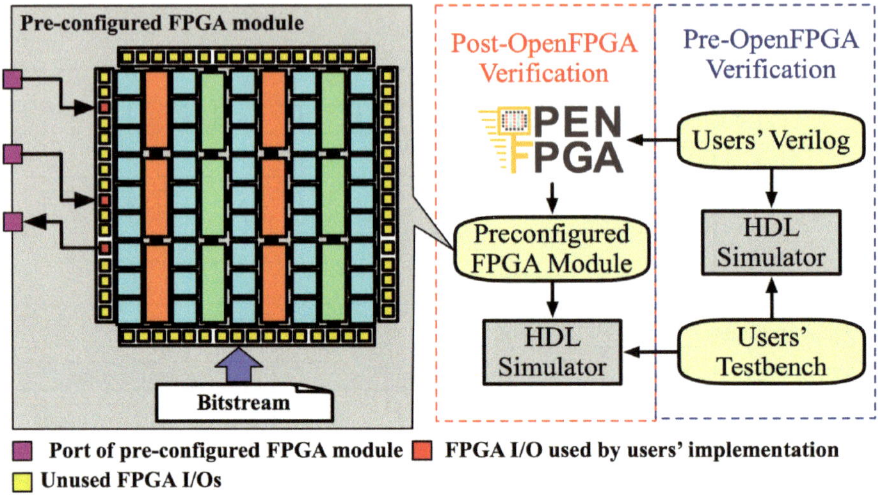

Fig. 7.17 Internal structure of a pre-configured FPGA module

7.5 Showcase

In this part, three FPGA fabrics produced by semi-custom EDA approaches are presented and then compared to a commercial baseline Stratix IV [24]:

1. a 20 × 20 homogeneous FPGA using a commercial 40 nm technology, built with standard cells only [8] (see layout in Fig. 7.3a);
2. a 20 × 20 homogeneous FPGA using a commercial 40 nm technology, built with standard cells only [18] (see layout in Fig. 7.18a);
3. a 32 × 32 heterogeneous FPGA using a commercial 12 nm technology, built with a mix of standard cells and custom cells [14] (see layout in Fig. 7.18b).

Note that through semi-custom approaches, the layout generation of the FPGA fabrics are within 24 h, but their architectures, technologies, and detailed methodologies are different. In all the FPGAs, each tile includes 10 *Logic Elements* (LEs) and a local routing architecture with 50% connectivity. The LE of homogeneous FPGAs consists of a 6-input fracturable LUT, a 4-input LUT, two 1-bit adders, and two flip-flops, which can operate in 6 different modes. The heterogeneous FPGA employs a simplified LE but without the 4-input LUT and also consists of a column of 512 Kb *Block RAMs* (BRAMs), generated by a foundry memory compiler. Full details about the showcased FPGA fabrics are listed in Table 7.5.

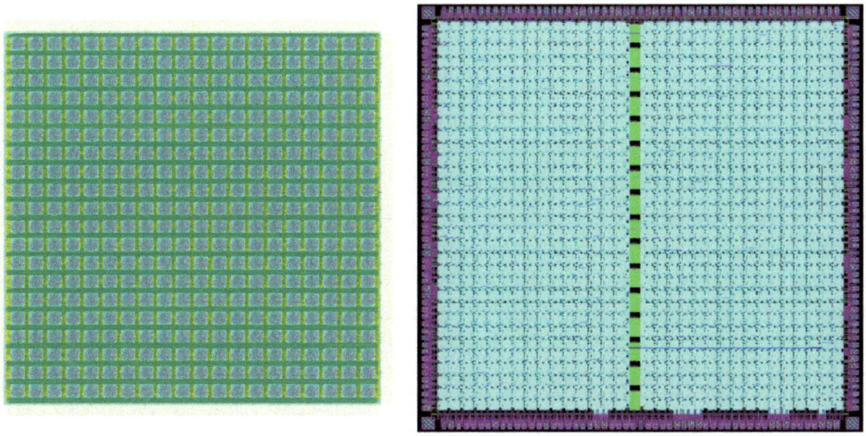

(a) Homogeneous (40nm) (b) Heterogeneous (12nm)

Fig. 7.18 Complete layout of FPGA fabrics

Table 7.5 Comparison on the FPGAs in Figs. 7.3a and 7.18

Resource/capacity	Standard homo [8]	Custom homo [18]	Standard hetero [14]
Array size	20 × 20	20 × 20	32 × 32
Tileable routing	×	×	✓
Fracturable 6-input LUTs	4 k [a]	4 k	9.92 k
4-input LUTs	N/A	8 k	N/A
1-bit full adder	8 k	8 k	19.84 k
Flip-flops	8k	8k	19.84 k
Block RAM	N/A	N/A	512 k bits
I/Os	N/A [b]	480	124
Routing channel width	300	300	200
Routing wires	87% L4	87% L4	L4
	13% L16	13% L16	
Fc_{in}	0.055	0.055	0.15
Routing multiplexer	tree-like	one/two-level	tree-like
Backend details	Standard homo [8]	Custom homo [18]	Standard hetero [14]
Tool	Cadence encounter v09.12	Cadence Innovus 19.1	Synopsys ICC2 2019.03
Layout area	16.89 mm^2	7 mm^2	9 mm^2
Flow type	Flatten	Two-step flatten	Hierarchical
Runtime (h)	20–24	24	12
Peak memory (GB)	64	60	215

[a] Each 6-input LUT contains 8 inputs
[b] Not reported

7.5.1 Methodologies

The homogeneous FPGA in [8] is generated by an in-house netlist generator based on VTR, while the rest of FPGA fabrics are generated by OpenFPGA [18]. Note that the netlists for the homogeneous FPGA in [8] were auto-generated in behavioral Verilog codes and optimized by Synopsys Design Compiler before physical design with a strategy to balance area and delay. The netlists auto-generated by OpenFPGA are technology mapped and directly used for physical design tools. Regarding circuit designs, the homogeneous FPGA in [8] and the heterogeneous FPGA in [14] is built with standard cells provided by a commercial 40nm technology, while the homogeneous FPGA in [18] adapts custom cells for routing multiplexers and configuration memory elements. Note that the homogeneous FPGA in [18] uses two-level structures for the multiplexers in *Connection Blocks* (CBs) and *Switch Blocks* (SBs) and local routing architecture, while one-level structure for those in LE. To guarantee high-performance, routing multiplexers are buffered at both inputs and outputs while LUTs are buffered at inputs, outputs, and every two intermediate stages.

The FPGA fabrics are implemented using three different physical design strategies. The homogeneous FPGA in [8] was implemented using a flatten backend flow with design constraints to force layout regularities. The homogeneous FPGA in [18] was implemented using a two-step backend flow where *Configurable Logic Blocks* (CLBs) are P&Red first and then instantiated at the top-level as hard macros. To leverage the symmetry of an FPGA fabric, the heterogeneous FPGA adopted a more hierarchical backend flow, where a library of hard macros for CLBs, CBs, and SBs is built and then assembled in the final layout. The hierarchical backend flow allows chip designers to optimize each hard macro with respect to the timing constraints generated by our tool with few combinational loops to be broken. Therefore, the heterogeneous FPGA is larger in array size, while its backend is 2× faster than the homogeneous. Commercial signoff tools are then used to ensure that all the fabrics are DRC-clean, and timing extraction is performed by using Synopsys PrimeTime.

7.5.2 Performance Evaluation

For a comprehensive analysis, the area, pin-to-pin delays, and the delays of the implemented benchmarks are considered when evaluating the FPGA fabrics. Table 7.6 compares the custom homogeneous FPGA in [18] to two baselines, a commercial Stratix-IV FPGA and the standard homogeneous FPGA in [8]. We believe it is a fair comparison since these FPGAs are similar in architecture and also implemented using 40nm technologies. The results prove the high value of using one-level and two-level multiplexing structures as well as an optimized cell library, which can improve the area by 42% and path delay by 30% when compared to a standard cell FPGA. Indeed, there are considerable gaps in area (60%) and path delays (20%) between the semi-custom-designed FPGAs and the full-custom-designed commercial FPGA.

Table 7.6 Area and delay comparison between [8, 14, 18] and Stratix IV

Generality	Standard homo [8]	Custom homo [18]	Standard hetero [14]	Stratix IV
Technology	40 nm	40 nm	40 nm	12nm
Cell Library	Standard	Custom[a]	Custom	Standard
Tile Area (μm^2)	30,625 (100%)	17,648 (-42%)	11,050 (-63%)	8,373 (-72%)
Path delay (ns)	Standard homo [8]	Custom homo [18]	Standard hetero [14]	Stratix IV
Process Corner	TT	SS	SS[b]	TT
6-LUT	0.5 (100%)	0.27 (−46%)	0.28 (−44%)	0.23 (−54%)
20-bit Adder[c]	1.63 (100%)	2.12 (+30%)	1.23 (-25%)	1.13 (−31%)
Local Routing[d]	0.27 (100%)	0.17 (−37%)	0.23 (-15%)	0.15 (−44%)
L4 track[e]	2.53 (100%)	0.82 (−67%)	0.59 (−76%)	0.75 (−70%)
Average	100%	−30%	−40%	(−50%)

[a]Use custom cells only in routing multiplexers and configuration chains
[b]The rest are standard cells. See details in [18]
[c]Consider the slow model in Quartus STA
[d]Local routing path starts from a BLE output and ends at a BLE input
[e]LX track: FF→length-X wire→Local Routing→LUT→FF

Even though there is an intrinsic PPA gap between standard-cell layouts and full-custom layouts, the performance gap can be reduced through a careful co-design between backend strategies and custom cell implementations [7].

For performance benchmarking, eight MCNC circuits are selected to fit all the 40nm FPGAs. Each benchmark is verified through the verification techniques in Sect. 7.4, using Mentor ModelSim and Synopsys Formality. Quartus 18.1.0 is used to implement the same benchmark set as the industry baseline, and the device model is set to the Stratix IV EP4S40G2F40C2. Figure 7.19a shows that FPGAs using custom cells is 2× slower on average than the Stratix IV. The gap comes from the hardware lags in performance, with an average of 20%. When critical paths consist of multiple routing paths listed in [Tab. 7.6], the delay difference will aggregate. The gap comes from sources:

1. the hardware lags in performance with an average of 10%. When critical paths consist of multiple routing paths listed in Table 7.6, the delay difference will aggregate. Therefore, the longer the critical path is, the larger the performance gap will be.

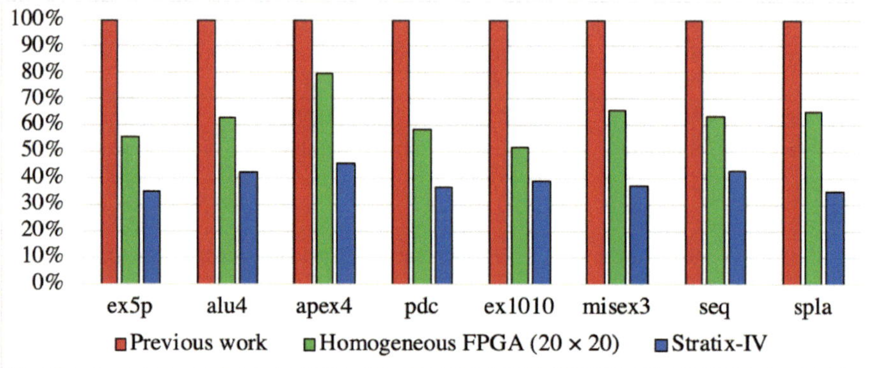

(a) Impact of custom cells and multiplexers on FPGAs at 40nm

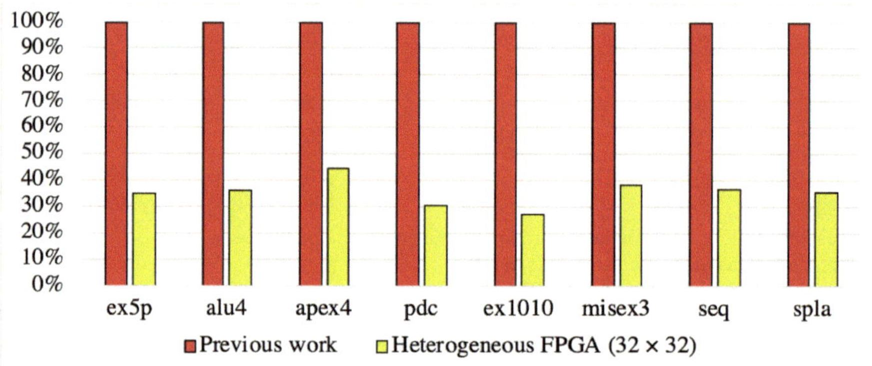

(b) Standard cell FPGAs scaling from 40nm to 12nm

Fig. 7.19 Delay comparison between OpenFPGA and [8] (marked as previous works) using selected MCNC benchmarks

2. previous studies have shown a large gap between VPR CAD algorithms and commercial counterparts [25]. The performance gap may be as large as 55% on average, fully shadowing any efficiency on hardware.

This indicates that developing efficient CAD algorithms that can match industry quality should be a frontier for the open-source FPGA research community.

We compare the heterogeneous FPGA in Fig. 7.18b to the homogeneous FPGA [8], as both FPGAs are implemented by standard cells and also similar in architecture while using different technologies. Our results show that using semi-custom design approaches, FPGA architectures can be portable between different technology nodes and benefit significant performance improvements. In Table 7.6, the 12 nm FPGA is 72% smaller in area and 50% faster in path than the 40nm baseline. In Fig. 7.19b, the heterogeneous FPGA is 3× faster on average in benchmark delays than the 40 nm baseline.

7.6 Summary and Trends

Semi-custom design approaches have become a warm research topics in recent years, as different design methodology than commercial state-of-the-art FPGAs that are built through full custom approaches.

To enable semi-custom design approaches, innovative EDA tools have been developed as an unified framework for netlist generation, testbench generation and bitstream generation. Due to the automation in modern EDA tools, development cycle of FPGA layouts as well as engineering effort can be remarkably reduced. However, the semi-custom design approach is in its infancy stage, as we see non-negligible PPA gaps against commercial FPGAs.

Since most of the EDA tools are accessible in open-source community, future researches may focus on performance improvement on the design methodology, e.g., physical design techniques. In addition, being tightly integrated to architecture exploration tools, the EDA tools enable fast prototyping for innovative FPGA architectures. In other words, architecture exploration can achieve realistic PPA evaluation in a short development cycle, and effectiveness of architecture enhancements can be validated through layout-level results in a short period, as compared the full-custom approach. Also, with the expansion in open-source community for FPGAs, novel EDA algorithms, e.g., packing, placement and routing, may be studied and validated through physical FPGA fabrics using semi-custom design approach. Previously, the validation of EDA algorithms is typically based on hypothetical FPGA fabrics and high-level analysis methods, which has been proven to be inaccurate.

In short, semi-custom design approaches have changed the cost function to design, evaluate, and produce new FPGA fabrics, stimulating many research opportunitie in novel FPGA architecture, efficient physical design techniques, and novel EDA algorithms.

References

1. S. Che, J. Li, J.W. Sheaffer, K. Skadron, J. Lach, Accelerating compute-intensive applications with GPUS and FPGAs, in *2008 Symposium on Application Specific Processors* (2008), pp. 101–107
2. C. Zhang, P. Li, G. Sun, Y. Guan, B. Xiao, J. Cong, Optimizing FPGA-based accelerator design for deep convolutional neural networks, in *Proceedings of the 2015 ACM/SIGDA International Symposium on Field-Programmable Gate Arrays*, ser. FPGA '15. (Association for Computing Machinery, New York, NY, USA, 2015), pp. 161–170. [Online]. Available: https://doi.org/10.1145/2684746.2689060
3. J. Cong, Z. Fang, M. Huang, L. Wang, D. Wu, CPU-FPGA coscheduling for big data applications. IEEE Design Test **35**(1), 16–22 (2018)
4. K. Neshatpour, H.M. Mokrani, A. Sasan, H. Ghasemzadeh, S. Rafatirad, H. Homayoun, Architectural considerations for FPGA acceleration of machine learning applications in mapreduce," in *Proceedings of the 18th International Conference on Embedded Computer Systems: Architectures, Modeling, and Simulation*, ser. SAMOS '18 (Association for Computing Machinery,

New York, NY, USA 2018), pp. 89–96. [Online]. Available: https://doi.org/10.1145/3229631.3229639
5. D. Greenhill, R. Ho, D. Lewis, H. Schmit, K.H. Chan, A. Tong, S. Atsatt, D. How, P. McElheny, K. Duwel, J. Schulz, D. Faulkner, G. Iyer, G. Chen, H.K. Phoon, H.W. Lim, W.-Y. Koay, T. Garibay, 3.3 a 14nm 1ghz FPGA with 2.5d transceiver integration, in *2017 IEEE International Solid-State Circuits Conference (ISSCC)* (2017), pp. 54–55
6. I. Kuon, A. Egier, J. Rose, Design, layout and verification of an FPGA using automated tools, in *Proceedings of the 2005 ACM/SIGDA 13th International Symposium on Field-Programmable Gate Arrays*, ser. FPGA '05 (Association for Computing Machinery, New York, NY, USA, 2005), pp. 215–226. [Online]. Available: https://doi.org/10.1145/1046192.1046220
7. Aken'Ova, V., Saleh, R., A "soft++" EFPGA physical design approach with case studies in 180 nm and 90 nm, in *IEEE Computer Society Annual Symposium on Emerging VLSI Technologies and Architectures (ISVLSI'06)* (2006), pp. 6
8. J.H. Kim, J.H. Anderson, Synthesizable FPGA fabrics targetable by the verilog-to-routing (VTR) CAD flow, in *2015 25th International Conference on Field Programmable Logic and Applications (FPL)* (2015), pp. 1–8
9. B. Grady, J.H. Anderson, Synthesizable heterogeneous FPGA fabrics, in *2018 International Conference on Field-Programmable Technology (FPT)* (2018), pp. 222–229
10. H.J. Liu, Archipelago - an open source FPGA with toolflow support (2014)
11. A. Li, D. Wentzlaff, Prga: an open-source FPGA research and prototyping framework, in *The 2021 ACM/SIGDA International Symposium on Field-Programmable Gate Arrays*, ser. FPGA '21. (Association for Computing Machinery, New York, NY, USA, 2021), pp. 127–137. [Online]. Available: https://doi.org/10.1145/3431920.3439294
12. D. Koch, N. Dao, B. Healy, J. Yu, A. Attwood, Fabulous: an embedded FPGA framework, in *The 2021 ACM/SIGDA International Symposium on Field-Programmable Gate Arrays*, ser. FPGA '21. (Association for Computing Machinery, New York, NY, USA, 2021), pp. 45–56. [Online]. Available: https://doi.org/10.1145/3431920.3439302
13. P. Mohan, O. Atli, O. Kibar, M. Zackriya, L. Pileggi, K. Mai, Top-down physical design of soft embedded FPGA fabrics, in *The 2021 ACM/SIGDA International Symposium on Field-Programmable Gate Arrays*, ser. FPGA '21. (Association for Computing Machinery, New York, NY, USA, 2021), pp. 1–10. [Online]. Available: https://doi.org/10.1145/3431920.3439297
14. X. Tang, E. Giacomin, B. Chauviere, A. Alacchi, P.-E. Gaillardon, OpenFPGA: an open-source framework for agile prototyping customizable FPGAs. IEEE Micro **40**(4), 41–48 (2020)
15. J. Luu, J. Goeders, M. Wainberg, A. Somerville, T. Yu, K. Nasartschuk, M. Nasr, S. Wang, T. Liu, N. Ahmed, K.B. Kent, J. Anderson, J. Rose, V. Betz, VTR 7.0: next generation architecture and cad system for FPGAs. ACM Trans. Reconfigurable Technol. Syst. **7**(2) (2014). [Online]. Available: https://doi.org/10.1145/2617593
16. K.E. Murray, O. Petelin, S. Zhong, J.M. Wang, M. Eldafrawy, J.-P. Legault, E. Sha, A.G. Graham, J. Wu, M.J.P. Walker, H. Zeng, P. Patros, J. Luu, K.B. Kent, V. Betz, VTR 8: high-performance cad and customizable FPGA architecture modelling. ACM Trans. Reconfigurable Technol. Syst. **13**(2) (2020). [Online]. Available: https://doi.org/10.1145/3388617
17. J. Luu, Architecture-aware packing and cad infrastructure for field-programmable gate arrays. Ph.D. dissertation, University of Toronto (2014)
18. X. Tang, E. Giacomin, A. Alacchi, B. Chauviere, P.-E. Gaillardon, OpenFPGA: an opensource framework enabling rapid prototyping of customizable FPGAs, in *2019 29th International Conference on Field Programmable Logic and Applications (FPL)*. (IEEE, 2019), pp. 367–374
19. V. to Routing, Verilog-to-routing documentation (2022) [Online]. Available: https://docs.verilogtorouting.org/en/latest/arch/
20. X. Tang, OpenFPGA documentation (2022). [Online]. Available: https://openfpga.readthedocs.io/en/master/
21. I. Kuon, R. Tessier, J. Rose (2008)
22. X. Tang, E. Giacomin, G. De Micheli, P.-E. Gaillardon, Circuit designs of high-performance and low-power rram-based multiplexers based on 4t(ransistor)1r(ram) programming structure. IEEE Trans. Circ. Syst. I: Regular Papers **64**(5), 1173–1186 (2017)

23. G. Gore, X. Tang, P.-E. Gaillardon, A scalable and robust hierarchical floorplanning to enable 24-hour prototyping for 100k-LUT FPGAs, in *Proceedings of the 2021 International Symposium on Physical Design*, ser. ISPD '21. (Association for Computing Machinery, New York, NY, USA, 2021), pp. 135–142. [Online]. Available: https://doi.org/10.1145/3439706.3447047
24. D. Lewis, E. Ahmed, D. Cashman, T. Vanderhoek, C. Lane, A. Lee, P. Pan, Architectural enhancements in Stratix-III™ and Stratix-IV™, in *Proceedings of the ACM/SIGDA International Symposium on Field Programmable Gate Arrays*, FPGA '09. (Association for Computing Machinery, New York, NY, USA, 2009), pp. 33–42. [Online]. Available: https://doi.org/10.1145/1508128.1508135
25. E. Hung, Mind the (synthesis) gap: examining where academic FPGA tools lag behind industry, in *2015 25th International Conference on Field Programmable Logic and Applications (FPL)* (2015), pp. 1–4

Part V
FPGA Application Design EDA

Chapter 8
High-Level Synthesis

Abstract *High-level synthesis* (HLS) is the process of compiling a software program into a digital circuit. This chapter provides a view into the HLS design flow and presents algorithms, tools, and methods to generate digital circuits from software descriptions. It details FPGA-oriented HLS techniques, discusses recent HLS advancements, and outlines the current challenges of HLS for FPGAs.

8.1 Overview

High-level synthesis (HLS) is the process of automatically compiling a high-level software program (e.g., in C or C++) into a hardware design [1, 2]. HLS aims to increase designer productivity by allowing a higher abstraction level that eases and shortens the hardware design process. Furthermore, it intends to make hardware design available to programmers without hardware design expertise (e.g., software developers who wish to benefit from hardware parallelism) [1].

A standard HLS software-to-hardware flow is outlined in Fig. 8.1. The HLS frontend is a typical software compiler that parses the input code and transforms it into an optimized intermediate compiler representation. The remainder of the flow is hardware-specific: the HLS tool schedules operations of the intermediate representation into clock cycles and determines the required resources to implement the complete circuit; the end result is the description of the circuit at RTL (e.g., VHDL, Verilog) level. The remainder of this section elaborates on this process.

8.1.1 From Software Program to Intermediate Representation

Like a software compiler, an HLS tool parses the input software code and performs syntax and type checks. It then transforms it into an *intermediate representation* (IR) that typically describes the program in a graph or assembly form.

K. Tu et al., *FPGA EDA*, https://doi.org/10.1007/978-981-99-7755-0_8

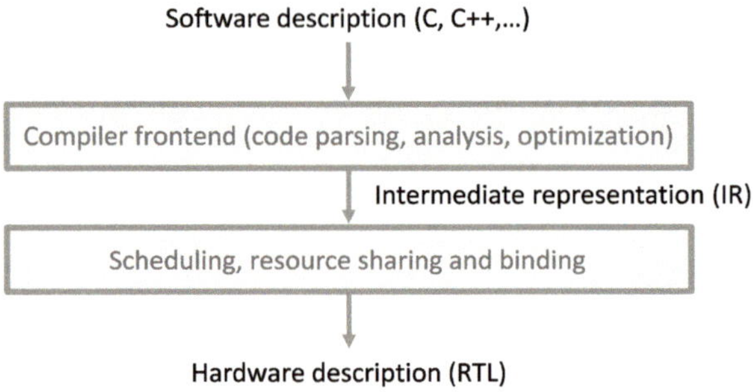

Fig. 8.1 High-level synthesis software-to-hardware flow

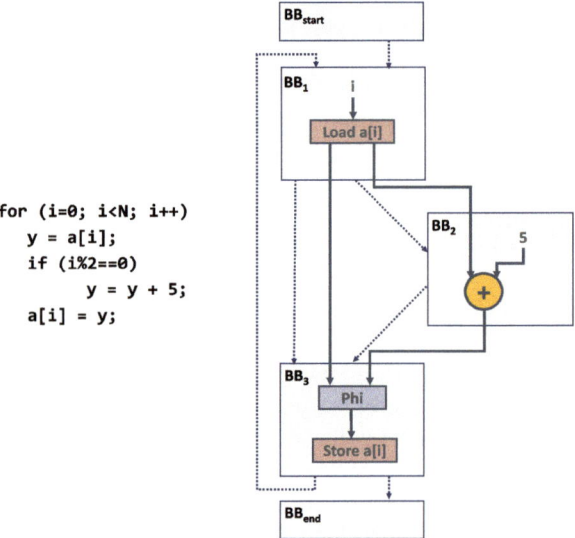

```
for (i=0; i<N; i++)
   y = a[i];
   if (i%2==0)
         y = y + 5;
   a[i] = y;
```

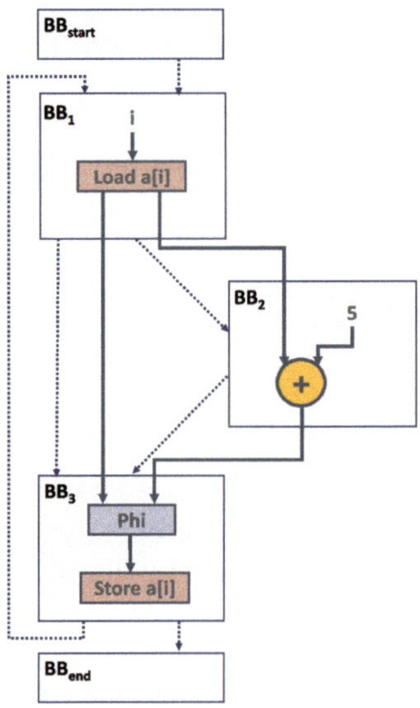

Fig. 8.2 An example of a compiler intermediate representation, organized into a control/dataflow graph

A standard way to represent producer-consumer relations among IR operations is a *dataflow* or *data dependence graph* (DFG). In a DFG, all program operations are represented as nodes and the data dependencies among them as edges. Conversely, a *control flow graph* (CFG) captures the control flow (i.e., conditional execution) of a program; it consists of *basic blocks* (BBs), connected by edges that represent the transfer of control from one BB to another. Internally, BBs are straight code sequences without any conditionals; all operations of a BB form a straight DFG that executes only when the condition to enter the BB has been determined [3].

Figure 8.2 shows an example of a program organized in a *control/data flow graph* (CDFG) which combines the concepts above in a hierarchical manner to describe all control and data dependencies of the original program. The control flow edges, connecting independent BBs, are shown in dashed. A portion of the datapath implementing the loop body is shown in the figure; edges between operations indicate data dependencies (i.e., producer-consumer relations). These concepts specify the execution order of particular operations: a producer operation must execute before its consumer; a BB of operations executes only after the condition to enter it through the appropriate control flow edge has been determined. Thus, they have a key role in scheduling, as we will discuss in the following sections.

The compiler frontend performs a variety of optimizations to make the IR as efficient as possible, thus enabling parallelism opportunities in the later stages of the HLS flow. For instance, it performs *code motion* to move computations from one CFG portion to another, *redundancy elimination* to remove the computation of values that have already been computed and can be used later in an unmodified form, and *tree balancing* to reduce long computational chains into compact structures. Additionally, it analyzes the code to support and enable later optimizations; for instance, *liveness analysis* determines the liveness of each variable and enables register allocation, *memory dependence analysis* enables the optimization of memory accesses and construction of efficient memory interfaces, and *loop unrolling* replicates the loop body for spatial hardware parallelism [3].

8.1.2 From Intermediate Representation to Hardware Design

Until this point, the program representation was *untimed*. A central task of HLS is to transform it into a *timed* representation that specifies the execution time of each event in the resulting hardware implementation. Therefore, the HLS tool *schedules* the operations of the IR into clock cycles while extracting as much parallelism as possible from the code; simultaneously, it decides on the position of registers to meet

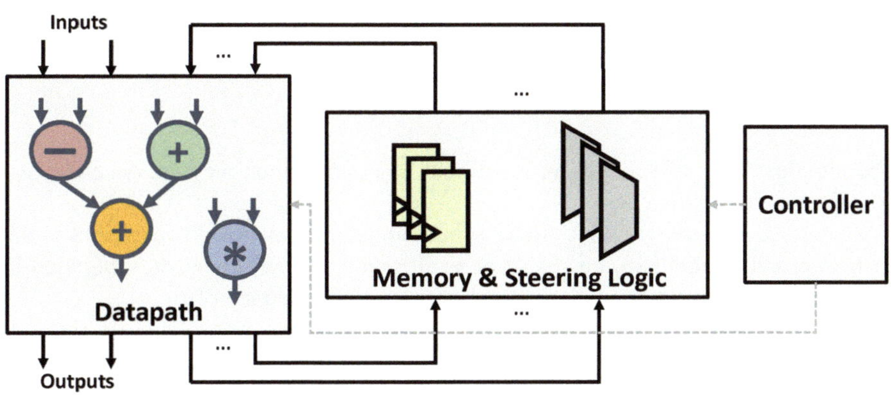

Fig. 8.3 An HLS-produced circuits is organized into a datapath of operations implementing the functionality of the input program, memory, and steering logic to send data to and from the datapath, and a controller that implements the schedule

the desired clock period target, maps operations onto the available FPGA resources, and defines the circuit interfaces that maximize the memory bandwidth [2].

The resulting circuit is organized as shown in Fig. 8.3:

1. The *datapath* contains functional units implementing the operations of the original code.
2. The *memory elements (i.e., registers)* store data items and the *steering and multiplexing logic* moves the data into the datapath and memories.
3. The *controller*, typically implemented as a finite state machine, dictates the operation schedule by producing enable signals for the registers and select signals for the multiplexers; it orchestrates the steering of data to and from the circuit (e.g., memory, input, and output ports) at appropriate times. Ultimately, the HLS compiler produces an RTL description of the circuit that can then be passed down to FPGA vendor tools for synthesis, placement, and routing [1].

8.2 Datapath Scheduling

Scheduling is the process of converting an untimed program representation into a timed representation by assigning each operation of a program to a time slot—typically, described in discrete time units, such as *clock cycles*. The duration of the clock cycle directly determines the *operating frequency* of the circuit and the total number of clock cycles determines the execution *latency*.

The HLS tool devises the operation schedule according to some optimization objective (e.g., minimizing the latency to achieve high performance); as mentioned in Sect. 8.1.2, it subsequently devises a controller that enforces this schedule by triggering operations at appropriate times. The scheduling process can be *unconstrained* or *constrained* by a variety of resource, timing, and latency constraints, which complexify the scheduling problem.

8.2.1 Unconstrained Scheduling

The simplest form of scheduling is without any constraints; we here describe two complementary approaches.

As soon as possible (ASAP). ASAP scheduling aims to schedule operations in the earliest possible time slot, i.e., as soon as all predecessors have been scheduled in some preceding time step, with the goal of minimizing latency [4].

Figure 8.5 shows an example of an ASAP schedule for the DFG of Fig. 8.4. Although not explicit in the figure, in the circuit implementation, each edge between operations will require a register whenever it crosses from one time step to another, to store the data that will be read on the following cycle. The resources (i.e., number of functional units) that the circuit implementation will require are bound by the maximal number of concurrent operations (i.e., operations scheduled in the same time step) of the same type, as they must execute on different functional units; operations

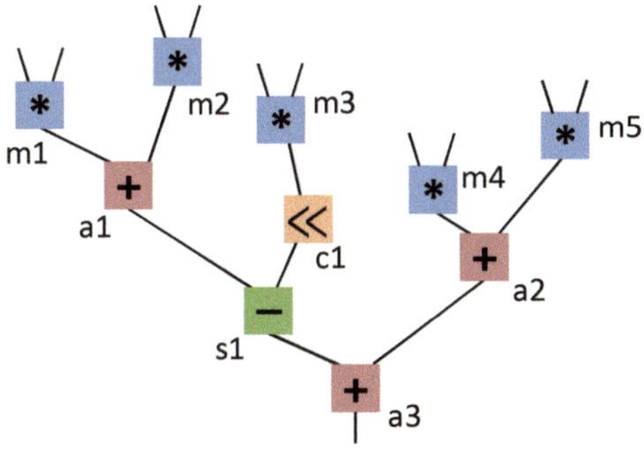

Fig. 8.4 A non-scheduled DFG of operations

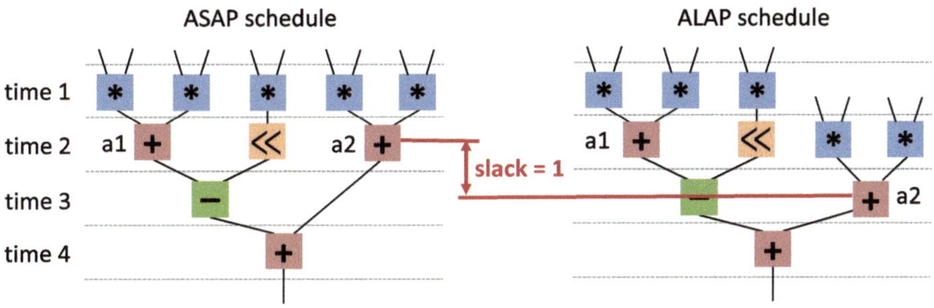

Fig. 8.5 ASAP and ALAP schedule of the DFG in Fig. 8.4

executing in different times can reuse the same functional unit, as we will discuss in Sect. 8.2.2.

As late as possible (ALAP). ALAP scheduling is complementary to ASAP: operations are scheduled as late as possible, starting from the sink of the graph and moving toward the earlier time steps; an operation is scheduled as soon as all of its successors have been scheduled [4].

Figure 8.5 contrasts the ALAP schedule with the ASAP schedule of the same graph. Both schedules achieve the best possible (i.e., minimal) latency; some operations are scheduled in the same time step in both schedules, whereas others are scheduled in a later step. The difference in operation start times between the ASAP and the ALAP schedule is referred to as *slack*. If the slack of an operation is greater than zero, it can be scheduled to another time slot without compromising the latency; a slack of zero indicates that the operation is on a latency-defining path and, thus, its movement would increase the latency. In this example, a_2 can be moved freely between time slots 2 and 3; however, a movement of a_1 would shift all succeeding operations and increase the latency to 5. The notion of slack is important when minimizing the resources under a latency constraint: it can be exploited to minimize the number of concurrent operations of the same type without a latency penalty.

$x_{op,t}$ → binary variable, indicating whether op starts in time t

Operation start:
$x_{m3,1} + x_{m3,2} = 1$ → Op m3 can be scheduled either in time 1 or in time 2

Sequencing:
$2x_{c1,2} + 3x_{c1,3} - x_{m3,1} - 2x_{m3,2} - 1 \geq 0$ → Op m3 must be scheduled earlier than op c1

Resource constraints:
$x_{m1,1} + x_{m2,1} + x_{m3,1} + x_{m4,1} + x_{m5,1} \leq 2$ → At most two muls can be scheduled in time 1

Fig. 8.6 A portion of an integer linear programming scheduling formulation for the example from Fig. 8.5

8.2.2 Constrained Scheduling

In real-life situations, scheduling can be constrained due to a variety of factors that impact the resulting schedule and its achievable latency. A common scheduling formulation accounts for a fixed number of available resources, thus requiring the latency and area to be traded off in different ways.

Integer linear programming (ILP). An exact scheduling problem is typically formulated as an ILP problem. The constraints are formulated as a system of linear constraints; the objective function minimizes latency under these constraints and the resulting integer values represent the clock cycle in which each operation needs to be scheduled [4].

Figure 8.6 shows examples of ILP constraints, formulated for the graph of Fig. 8.4 and assuming a latency bound of 5. In the equations, $x_{\text{op},t}$ is a binary variable indicating whether operation *op* starts in time *t*. The constraints on the operation start times specify that each operation can start only once (in the first equation, m_3 can start in time 1 or in time 2). The sequencing constraints indicate the timing relations between different operations (in the second equation, m_3 must start before its successor c_1). The resource constraints specify the maximal number of units of the same type in every time step (in the third equation, 2 multipliers). These constraints can be used with different ILP objective functions—for instance, to minimize latency, the objective function minimizes the start times of all operations.

Scheduling under resource constraints is an NP-hard problem; thus, in addition to exact algorithms, there are many approximate ways to identify an acceptable solution efficiently for complex graphs.

List scheduling. The idea of list scheduling is to prioritize the scheduling of certain operations based on an urgency metric. Typical examples include the length of the path from the operation to the sink of the graph (where a longer path corresponds to higher urgency) or slack (where a lower slack corresponds to higher urgency).

Figure 8.7 shows a schedule obtained through list scheduling by prioritizing operations on the longest path to the sink and a resource constraint of 2 multipliers (the first two scheduling steps are indicated below the schedule), contrasted with the same

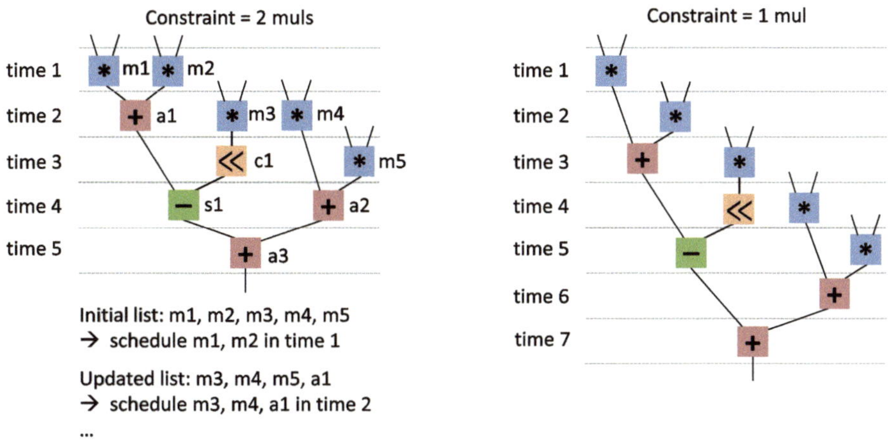

Fig. 8.7 List scheduling given a resource constraint of 2 multipliers (left) and 1 multiplier (right)

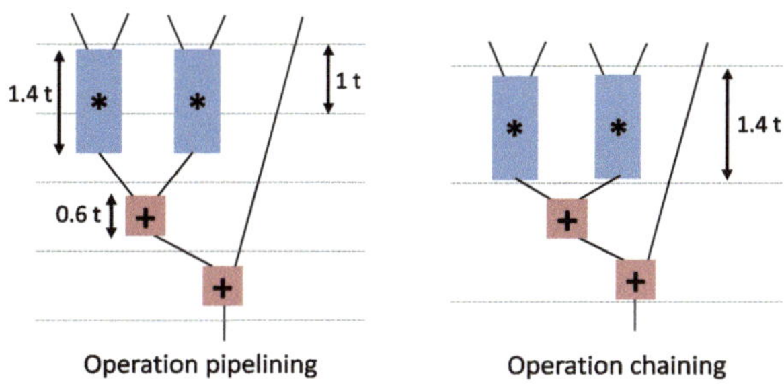

Fig. 8.8 Optimizing timing through operation pipelining (left) and chaining (right)

scheduling strategy with a constraint of 1 multiplier. Tightening the resource constraint comes at a latency penalty, which is a typical area-performance trade-off of constraint scheduling.

List scheduling is a heuristic approach: the information available in each particular step provides no information on further steps and the potential conflicts that high-priority operations will encounter. Thus, minimal latency is not guaranteed, but its low complexity of *O(n)* makes it an attractive solution for complex applications [4].

8.2.3 *Timing Optimizations*

Another timing aspect that HLS scheduling typically exploits is the ability to adjust the clock period of the circuit by adding or removing registers and trading off latency and the clock period in different ways. As shown in Fig. 8.8, *pipelining* inserts registers to break operations or paths into multiple time slots to reduce the circuit's

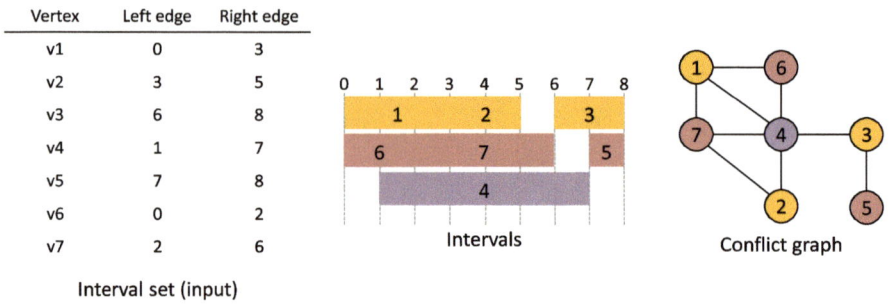

Vertex	Left edge	Right edge
v1	0	3
v2	3	5
v3	6	8
v4	1	7
v5	7	8
v6	0	2
v7	2	6

Interval set (input)

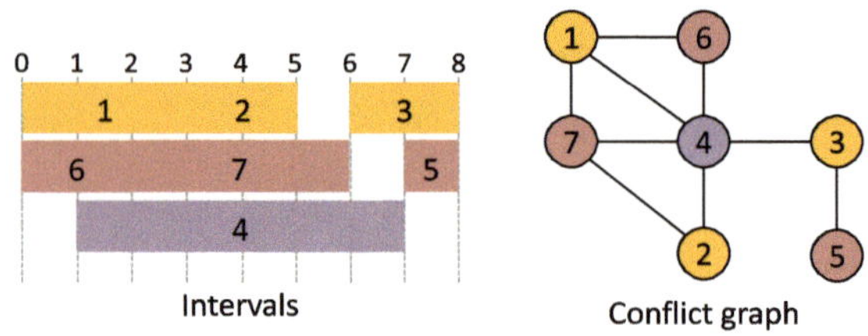

Fig. 8.9 Left-edge algorithm for resource sharing

critical path, at a possible latency expense. Conversely, *operation chaining* fits multiple operations into a single clock cycle to execute combinationally; it saves registers and latency on combinational paths that are not the critical path of the circuit.

These optimizations are, thus, also included in modern HLS scheduling formulations as timing constraints, on top of the resource constraints discussed in the previous section.

8.2.4 Resource Binding and Sharing

Resource binding is the process of mapping operations of the program to physical resources. It can be accompanied by *resource sharing* to assign a single resource to multiple non-concurrent operations [4]. Binding and sharing can be applied on a scheduled or non-scheduled graph; we here illustrate the problem on a scheduled graph.

To identify which operations are compatible to be implemented on the same resource, operations can be represented using *compatibility* and *conflict* graphs. In a compatibility graph, the edges denote compatible, i.e., non-concurrent operation pairs, that can be implemented on the same resource. The sharing problem can then be solved by *clique partitioning*, where each resulting clique corresponds to a resource instance; optimal sharing is achieved by partitioning into a minimal clique number. The dual problem is to reason about the conflict graph. Edges denote conflicting operations and the sharing problem can be solved by *graph coloring*; optimal sharing is achieved by coloring with a minimal color number, where each color represents the resource instance (see right of Fig. 8.9).

In general, vertex coloring is an intractable problem. Yet, if the graph is represented as an *interval graph*, the coloring can be achieved in polynomial time. This is the intuition behind the *left-edge algorithm* [4] that formulates the sharing problem on an interval graph. The input to the algorithm is a set of execution intervals for each operation. The rationale is to sort the intervals in a list by the left edge (i.e., based on their earliest possible start times) and assign non-overlapping intervals to a single color; when all intervals of a color are exhausted, a new color is introduced and the procedure repeated.

An example is shown in Fig. 8.9. Vertex v_1 is assigned the first color; it is followed by v_6 and v_4 that overlap with the interval of v_1 and, thus, require new colors. All other vertices can be assigned to existing intervals, resulting in a total of three colors (i.e., three functional units, each of them executing the operations of a single color).

Although we here discussed sharing and binding of functional units, the same applies to registers, memories, buses, and other resource types; state-of-the-art binding and scheduling HLS formulations consider all these aspects.

8.3 Extracting Parallelism Through HLS Scheduling

In this section, we discuss the state-of-the-art HLS scheduling algorithms for FPGAs. We present the concept of *system of difference constraints* (SDC) modulo scheduling that HLS tools today rely on; we outline polyhedral techniques for memory and loop optimizations. We then discuss the inability of these techniques to handle irregular behavior and more recent solutions to overcome these limitations.

8.3.1 SDC-Based Modulo Scheduling

The techniques of Sect. 8.2 minimize the latency of a single datapath, but they are not sufficient to extract parallelism when datapaths repeat (e.g., in a loop execution): simply executing one iteration after another in a sequential way would result in underutilized datapath resources and low performance, as shown on the left of Fig. 8.10: the total latency corresponds to the sum of latencies of individual datapaths, $N \cdot Lat$, where N is the number of iterations and Lat the latency of a single datapath.

Loop pipelining is one of the main performance optimization techniques in HLS—it allows the overlapping of loop iterations such that the datapath is used in the best possible way while honoring all data, control, and memory dependencies of the program. Pipelining originates from *modulo scheduling* techniques for *Very Long*

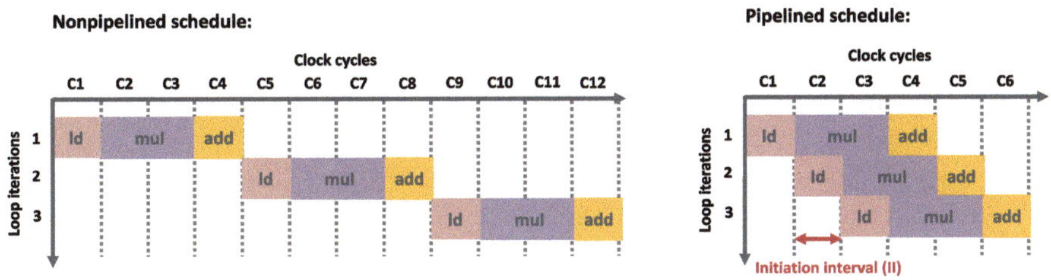


```
for (i=0; i<n; i++) {
        acc += x[i] * c[n-i];
}
```

Fig. 8.10 A non-pipelined (left) and a perfectly pipelined (right) schedule with an initiation interval of 1

Instruction Word (VLIW) processors [5], that aim to restructure the code to exploit instruction-level parallelism among loop iterations. As in the case of VLIWs and as discussed before, it is up to the HLS compiler to devise the schedule and create a controller (i.e., a finite state machine) that triggers operations according to this schedule.

A pipeline is characterized by its *initiation interval* (II), defined as the number of clock cycles between consecutive loop iterations. The best possible II is equal to 1 and indicates that a new iteration starts on every consecutive clock cycle (this is the case for the schedule on the right of Fig. 8.10). The total pipeline latency is now $(N - 1) \cdot \text{II} + Lat$. The II increases in the presence of data, memory, or control dependencies between iterations, which postpone the start time of the next iteration and thus lower performance. Similarly, if a dependency is undeterminable at compile time, the HLS tool must assume its presence and increase the II accordingly; we will discuss this scenario in Sect. 8.3.3.

State-of-the-art commercial and academic HLS solutions today rely on SDC modulo scheduling [6–8] to achieve high-throughput pipelines. The idea is to describe all scheduling constraints as a system of difference constraints in the form $x - y \leq b$, where x and y are integer variables and b is a constant, and formulate a linear programming problem that minimizes the II under these constraints. Such a formulation supports a wide variety of constraints, such as data dependencies among operations, control dependencies between BBs, frequency, latency, and various resource constraints. The SDC scheduling problem is typically solved iteratively: the HLS tool attempts to find a solution for the desired II; in case it is not found, the II is incremented and the scheduling procedure is repeated [7].

8.3.2 Polyhedral Analysis and Optimization

HLS tools must handle the scheduling of complex programs; thus, they need to describe program features in a compact, parametric, and general way.

Polyhedral analysis is a powerful compiler technique for describing program features such as loop properties (e.g., loop bounds, iterations, and strides) and memory accesses (e.g., memory access patterns and dependencies). It is used to reason about *Static Control Parts* (SCoPs) of the program, i.e., regions in which all control flow decisions and memory accesses are known at compile time. Within a SCoP, all loops and memory accesses can be described using integer polyhedra [9, 10].

Figure 8.11 shows examples of loops that are SCoPs: they have affine expressions in induction variables and parameters for loop bounds, control flow decisions, and memory accesses. The loops at the bottom of the figure cannot be described as SCoPs nor optimized with polyhedral analysis. In addition to loop properties, polyhedral techniques describe the memory access pattern of each load and store instruction, as illustrated in Fig. 8.12. Together with the schedule, this information is key to identify all read-after-write (RAW), write-after-read (WAR), and write-after-write (WAW) dependencies of the program.

Loops that are SCoPs

```
for (i=0; i<N; i++)
    a[i]=b[i]+5;
```

```
for (i=0; i<M; i++)
  for (j=0; j<N; j++)
    for (k=min(i,j); k<M; k++)
      a[i][j]=a[i][k]-b[j][k];
```

```
for (i=0; i<1000; i++)
  if (i%2==0)
    a[i]=0;
```

Loops that are not SCoPs

```
for (i=0; i<N; i*i)
    a[i]=b[i]+5;
```

Non-affine loop increment

```
for (i=0; i<1000; i++)
    a[b[i]]=c[b[i]];
```

Indirect (statically undeterminable) memory accesses

```
for (i=0; i<1000; i++)
  if (a[i]%2==0)
    a[i] = 0;
```

Data-dependent control flow

Fig. 8.11 Loops that can be described as SCoPs and optimized with polyhedral analysis (top), and loops that are not SCoPs (bottom)

```
for (i=0; i<N; i++)
   for (j=0; j<N; j++)
      A[2i+5][3j] = i*j;
```

Iteration domain: $\{i, j \mid 0 \le i < N, i \le j < N\}$

Access pattern of A: $\{2i + 5, 3j \mid i, j \in Z\}$

Accessed memory locations: $\{i, j \mid 5 \le i < 2N + 4, 0 \le j < 3N\}$

Fig. 8.12 A parametric description of the iterations and memory accesses of a SCoP

Program transformations. The goal of polyhedral optimization is to simplify the program and expose parallelism by reordering loops and loop iterations and changing the program's memory access patterns. The optimizations are performed by applying linear transformations on SCoPs such that the program semantics are preserved: independent loop iterations can be reordered and restructured, but those containing dependent references (e.g., memory dependencies) must be executed in the same order as in the original program [9, 10].

Polyhedral techniques are useful for optimizing HLS programs with statically determinable properties, as they can uncover new parallelism opportunities [10] that scheduling algorithms such as SDC scheduling (see Sect. 8.3.1) can exploit. However, they cannot be applied in general-purpose programs with irregular behaviors that cannot be captured by SCoPs.

8.3.3 Dynamic Scheduling

The techniques from the previous sections rely on the HLS compiler to devise the best possible schedule; while this approach is effective when critical information on program execution and behavior is available *at compile time*, it fails in cases with statically undeterminable memory accesses, variable operation latencies, and unpredictable control flow. In such situations, the HLS tool must assume the worst-

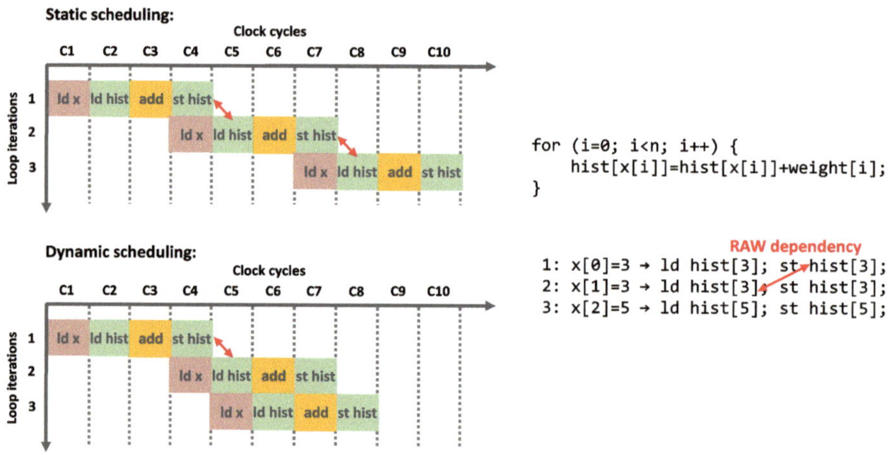

Fig. 8.13 A static schedule (top), achieved by a standard HLS tool, and a dynamic schedule (bottom), which achieves higher parallelism by resolving memory accesses dynamically at circuit runtime. The schedules are realized by the circuits shown in Fig. 8.14

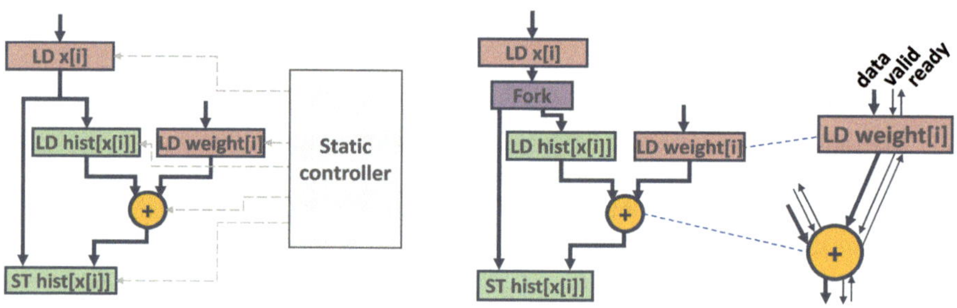

Fig. 8.14 A portion of a statically scheduled circuit (left) and a dynamically scheduled circuit (right), implementing the behavior of the code in Fig. 8.13

case scenario and devise the schedule accordingly, which often results in suboptimal throughput and low performance.

An example of one such situation is shown in Fig. 8.13. The code in the figure has indirect memory accesses to array *hist*; depending on the values of x, there may or may not be a RAW dependency between a load and a store from a previous iteration. A standard HLS tool must assume the presence of a dependency, devise the appropriate scheduling constraint (similar to the sequencing constraint in Fig. 8.6), and create a conservative schedule (top of the figure), assuming that a dependency is always present.

The most general way to avoid the limitations of static scheduling is to forgo operation triggering through a pre-planned, statically scheduled controller (as shown on the left of Fig. 8.14) that dictates the exact execution time of each operation, but to make scheduling decisions as the circuit runs: as soon as all conditions for execution are satisfied (e.g., the operands are available or critical control decisions are resolved), an operation starts. *Dataflow circuits* [11–13] are a natural method to realize such behavior. They are built out of units that implement *latency-insensitivity* by communicating with their predecessors and successors through pairs of handshake control signals, which indicate the availability of a new piece of data from the source unit and the readiness of the target unit to accept it (as illustrated on the right of Fig. 8.14). The data is propagated from unit to unit as soon as memory and control dependencies allow it and stalled by the handshaking mechanism otherwise, thus effectively devising a *dynamic schedule* at circuit runtime. Such a schedule is shown at the bottom of Fig. 8.13: the pipeline is stalled only when a dependency actually exists, otherwise, the loads and stores to array *hist* may execute out of order for performance benefits.

Several works generate dataflow circuits from functional and imperative software program representations [14–16]. The most recent effort in the context of HLS for FPGAs is Dynamatic [16, 17], a complete and open-source HLS compiler that produces high-throughput dataflow circuits from C/C++ code. It incorporates features and compiler transformations to make dataflow circuits truly competitive in the context of modern HLS. The ability to adapt the schedule at runtime offers completely new optimization opportunities: memory dependencies can be resolved at runtime and key control decisions can be speculated on, just like in superscalar processors. Thus, dynamic HLS shows significant speedups when contrasted to state-of-the-art HLS tools [18, 19].

8.3.3.1 Pipelining and Resource Sharing in the Absence of a Static Schedule

Dataflow circuits must benefit from the same performance and area optimization opportunities as their statically scheduled counterparts; yet, classic scheduling and sharing algorithms described in Sect. 8.2 are not applicable in this context.

In contrast to devising a predetermined pipeline with a fixed II (e.g., by SDC modulo scheduling, as described in Sect. 8.3.1), the performance of dataflow circuits can be optimized via *slack matching*: inserting pipeline buffers (i.e., FIFOs) of appropriate sizes to prevent stalls and increase parallelism [20, 21], as shown on the left of Fig. 8.15. Slack matching can be combined with frequency optimization to achieve high-throughput synchronous dataflow circuits that honor the required clock period constraint [22, 23]. Similarly, in dataflow circuits, the sharing suitability of operations depends on runtime decisions and schedule adaptations—classic resource sharing techniques that rely on compile time concurrency information (see Sect. 8.2.4) are therefore not applicable. Thus, instead of reasoning about the exact execution times of each operator, dynamic HLS relies on the information on average unit utilization in the steady state of the system and determines what to share accordingly [24]; a sharing implementation of a dataflow unit is shown on the right of Fig. 8.15. These optimizations are key to making dynamic HLS performance- and resource-competitive with static HLS designs.

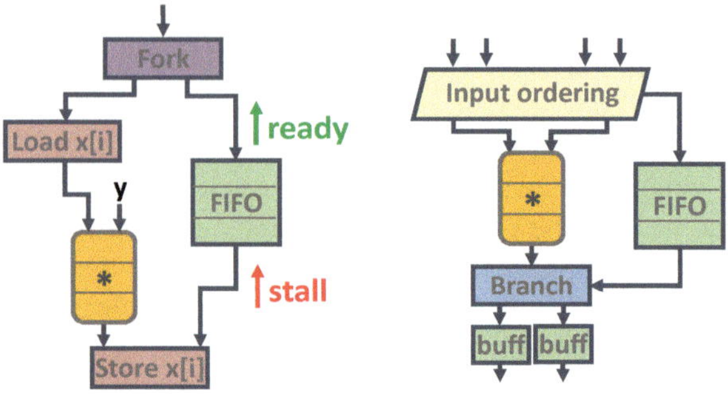

Fig. 8.15 Pipelining dataflow circuits by inserting FIFOs (left) and a mechanism for sharing dataflow units (right)

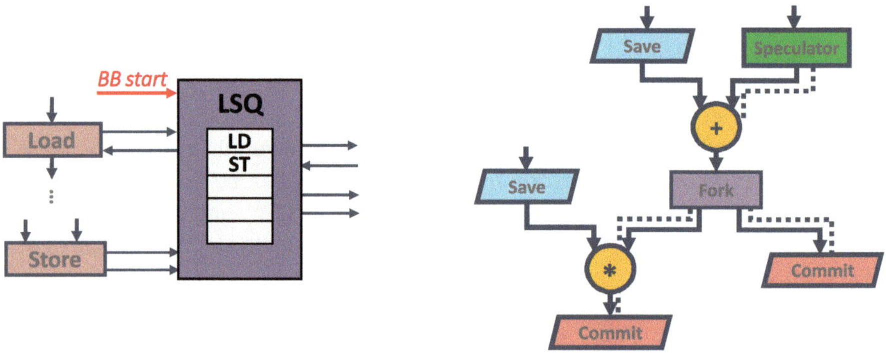

Fig. 8.16 A load-store queue for dynamic memory dependency resolution (left) and a distributed speculation mechanism for dataflow circuits (right)

8.3.3.2 Dynamic Scheduling and Irregular Memory Accesses

When memory dependencies are statically unknown, standard HLS must assume the presence of the dependency—in terms of the scheduling algorithms above, a data dependency constraint conservatively dictates that the two accesses must be sequentialized. In contrast, dataflow circuits determine the presence or absence of a dependency during runtime using *load-store queues* (LSQs) [25–27], such as the one on the left of (Fig. 8.16). They compare the possibly conflicting addresses at runtime and enforce the access order *only* when they are dependent (e.g., in the example of Fig. 8.13, the LSQ postpones the load of the second iteration until the previous store, that targets the same address, completes); if the accesses are independent, the LSQ allows them to execute out of order (this is the case for the load of the third iteration, that executes before the preceding store). This scheduling flexibility and its performance benefits are impossible in static HLS.

8.3.3.3 Dynamic Scheduling and Speculative Execution

Speculation is a classic superscalar processor feature that can significantly improve the performance of loops where the loop condition takes a long time to compute by tentatively starting a new loop iteration before the loop condition is known. In static HLS, this optimization is limited to only trivial cases and otherwise hindered by the inability of the static schedule to revert the execution to a prior state in case of a misspeculation. Instead, dataflow circuits support generic forms of speculation [28]: speculative data travels through the circuit and dedicated components implement a distributed squash-and-replay mechanism (as shown on the right of Fig. 8.16), conceptually similar to that of superscalar processors, which achieves high parallelism in control-dominated applications.

8.3.3.4 VLIWs Versus Superscalars

Dynamic scheduling is in strong contrast to the strategy of Sect. 8.3.1 and in direct analogy to the contrast of VLIW and superscalar processor scheduling: In VLIWs, it is up to the compiler to devise the fixed schedule, which avoids the need to perform dependency checks at runtime (as the schedule guarantees that they are honored) and results in simpler hardware implementations [5, 29]. In contrast, superscalar processors [30] rely on more complex hardware mechanisms to resolve memory and control dependencies at runtime as well as to speculate on critical decisions; this flexibility makes them applicable to a wider variety of situations, which is why they are the generally accepted solution for general-purpose software applications. The situation is the same with static and dynamic scheduling in HLS: static HLS achieves great parallelism in particular application classes, where techniques such as polyhedral analysis and SDC modulo scheduling are successful, but irregular software programs require the flexibility of dynamic scheduling [18].

8.4 Current Status and Outlook

In this section, we provide an overview of current trends of HLS for FPGAs. We discuss typical usages of HLS and active open-source HLS frameworks, and outline some of the challenges that modern HLS is facing.

8.4.1 HLS Frameworks

In this section, we provide an overview of recent HLS frameworks targeting FPGAs.

C and OpenCL-based HLS frameworks. Apart from commercial HLS flows for FPGAs, such as AMD (formerly Xilinx) Vitis HLS [31] and Intel HLS [32], numerous open-source HLS projects are under active development. LegUp HLS [33] was originally developed as a complete open-source HLS flow, supporting C++ as well as task-oriented language constructs (e.g., OpenMP and Pthreads) [34]; it has recently been acquired by Microchip and is now closed-source. Bambu [35] is an open-source HLS research framework that supports a variety of FPGA backends and provides support for verification and debugging. Dynamatic [17] is an open-source HLS tool that produces dynamically scheduled dataflow circuits from C/C++ and supports features such as dynamic memory dependency resolution and speculation. DASS [36] has been developed on top of Dynamatic and Vitis HLS to combine the benefits of static and dynamic scheduling.

Compiler infrastructures. A majority of recent HLS flows rely on the well-established LLVM [37] compiler as a frontend to obtain an optimized intermediate representation from C/C++. LLVM provides a single IR describing the program as a CDFG, as described in Sect. 8.1.1; this serves as a starting point to implement either static or dynamic scheduling. Recently, MLIR [38] emerged as an alternative to LLVM; its compiler infrastructure allows the definition and composition of multiple IRs (also referred to as *dialects*), thus providing modularity and extensibility at different levels of abstraction. The CIRCT project [39] leverages MLIR to provide a variety of hardware-oriented features and abstractions; it incorporates some of the main transformations of Dynamatic, new IRs for hardware (e.g., Calyx [40]), and supports HLS code transformations (e.g., ScaleHLS, [41]). All this, together with a C-based frontend (i.e., Polygeist [42]), will likely serve as a basis for the development of future open-source HLS flows.

Domain-Specific Languages for HLS. Many HLS efforts explore domain-specific languages as an HLS frontend to raise the level of abstraction, increase productivity, and ease the expression of particular domain-specific constructs [43]. Popular DSLs target domains where HLS is successful, such as image processing (e.g., Halide [44–46]), machine learning (e.g., HeteroCL [47]), and streaming applications (e.g., Spatial [48]). Most DSLs use an existing C-based HLS flow as a backend and thus ultimately rely on the HLS techniques described in this chapter.

8.4.2 HLS Code Restructuring and Annotations

Thanks to the HLS frameworks and languages above, HLS is gaining popularity in domains such as machine learning, image processing, graph processing, video transcoding, and networking [49]. However, HLS is still facing critical adoption challenges due to the difficulties of extracting the desired levels of performance: despite the raised programming abstractions, HLS programmers still need to restructure the code and annotate it with pragmas to guide the HLS tool in achieving good parallelism and the desired hardware characteristics. This typically requires significant hardware design expertise and makes HLS unavailable to non-expert users [49, 50].

Consider the example in Fig. 8.17, illustrating a naively written code to add a set of integers held in external memory and store back the result, compared to a

```
void(int* mem){
    mem[512] = 0;
    for(int i=0; i<512; i++)
        mem[512] += mem[i];
}
```

```
// Memory port Port = 16 * sizeof(int)
#define Chunk (sizeof(Port)/sizeof(int))
#define Count (512/ChunkSize)
// Maximize data width from memory
void(Port * mem){
    // Local buffer and burst access
    Port buff[Count];
    memcpy(buff, mem, Count);
    // Local variable for accumulation
    int sum=0;
    for(int i=1; i<Count; i++){
    // Directives for parallelism
        #pragma PIPELINE
        for(int j=0; j<Chunk; j++){
            #pragma UNROLL
            sum+=(int)(buff[i]>>j*sizeof(int)*8);
        }
    }
    mem[512]=sum;
}
```

Fig. 8.17 HLS code restructuring to achieve high parallelism. Reproduced from George et al. [43]

restructured code achieving significantly better performance: the restructured code accounts for the data widths, memory interface communication, and other architectural aspects, thus allowing the HLS scheduler to exploit the parallelism available in the computation. This code writing style is not necessarily accessible to software programmers who do not have knowledge of the underlying architecture. Although several works attempt to automate the pragma insertion process [51, 52] and despite the ability of the DSLs to hide many hardware-oriented details, the challenges of HLS programming are still one of the main factors preventing its broad usage.

8.4.3 Design Space Exploration

All the scheduling possibilities and constraints discussed in the previous sections, as well as the large design space achievable by different restructurings and annotations, create a complex design space with a variety of non-trivial design options: loops can be pipelined with different initiation intervals and unrolled with different factors, resource constraints can be formulated for different FPGA resource types (e.g., DSPs, BRAMs, etc.); the design can be optimized for throughput or latency, as well as tuned to different frequencies. This forms a multi-objective optimization problem that aims to minimize a set of, possibly conflicting, design parameters; the result is a set of points forming a Pareto frontier [53].

Due to the large search space, it is difficult to evaluate design quality and to understand whether the Pareto-optimal points have actually been found or approached. Secondly, the design space exploration itself requires the evaluation of particular points to continue the exploration in the appropriate direction. Some approaches synthesize each point with HLS and evaluate it on the fly (at the expense of long runtimes), whereas others build analytical models to estimate the area and performance (which may be less accurate, but faster) [53, 54]. Finally, different HLS tools have different implementation strategies and design spaces, so it is challenging to directly apply the notions of one tool's DSE to the other [54]. With a plethora of new HLS techniques and relevant metrics arising, DSE will certainly increase in complexity—the ability to handle it efficiently will be key to navigate through this increasingly complex design space.

8.4.4 Functional and Formal Verification in HLS

HLS tools typically rely on functional verification of a particular circuit through hardware simulation and software/hardware cosimulation [55]. However, performing exhaustive hardware simulations may become unfeasible or extremely time-consuming as designs increase in complexity. Furthermore, the lack of formal proof on the correctness of particular compilation steps and the resulting hardware modules prevents the adoption of HLS in domains where design iterations are significantly more expensive [56].

Recent efforts aim to formally verify the HLS process [57, 58]. Others employ formal methods to optimize HLS-produced circuits: Cheng et al. use an SMT-based solver to improve the memory arbitration in HLS [59]. Geilen et al. employ model checking for buffering coarse-grain dataflow graphs [60]. Xu et al. use BDD-based reachability analysis [61] and induction [62] to prove particular behavioral properties of HLS-produced circuits and use them to improve their hardware implementation. Such formal methods are key to comprehensively reason about HLS transformations and the resulting circuits.

8.4.5 Frequency Estimates in HLS

A key task of HLS scheduling is to break combinational paths with registers and ensure that the circuit meets the target clock frequency, as discussed in Sect. 8.2.3. Yet, when placing registers, HLS typically relies on pre-characterized timing information [63] which fails to account for the effects of FPGA synthesis, placement, and routing, causing several undesired effects:

1. The overestimated unit latencies may cause conservative pipelining and unneeded resource overheads (due to the redundant register placements and the prevention of logic optimizations across register-separated pipeline stages).
2. The same conservative pipelining may unnecessarily decrease parallelism and, thus, performance, if a register is redundantly placed on a throughput-critical cycle.

3. During placement and routing, the backend of the FPGA flow can introduce delay variations caused by interconnect delays that are difficult to estimate and cause frequency discrepancies from the target [49].

Recent works extend HLS scheduling formulations, such as the ones in Sect. 8.3, with physical design objectives; they aim to make HLS optimizations aware of the physical layout of the FPGA by including LUT mapping information into the scheduling problem [63–65] and estimating routing congestion and interconnect delays [66, 67]. This information is critical to advance HLS design quality and make them comparable to hand-optimized RTL [49, 50].

References

1. M. Hutton, V. Betz, J. Anderson, FPGA synthesis and physical design, in *Electronic Design Automation for IC Implementation, Circuit Design, and Process Technology* (CRC Press, 2017), pp. 395–436
2. R. Kastner, J. Matai, S. Neuendorffer, Parallel programming for FPGAs (2018). ArXiv e-prints arXiv:1805.03648
3. L. Torczon, K. Cooper, *Engineering a Compiler*, 2nd ed. (Morgan Kaufmann, 2011)
4. G. De Micheli, *Synthesis and Optimization of Digital Circuits* (McGraw-Hill, 1994)
5. B.R. Rau, Iterative modulo scheduling. Int. J. Parallel Programm. **24**(1), 3–64 (1996)
6. J. Cong, Z. Zhang, An efficient and versatile scheduling algorithm based on SDC formulation, in *Proceedings of the 43rd Design Automation Conference* (San Francisco, CA, 2006) pp. 433–438
7. Z. Zhang, B. Liu, SDC-based modulo scheduling for pipeline synthesis, in *Proceedings of the 32nd International Conference on Computer-Aided Design* (San Jose, CA, 2013), pp. 211–218.
8. A. Canis, S.D. Brown, J.H. Anderson, Modulo SDC scheduling with recurrence minimization in high-level synthesis, in *Proceedings of the 23rd International Conference on Field-Programmable Logic and Applications* (Munich, 2014), pp. 1–8
9. L. Pouchet, P. Zhang, P. Sadayappan, J. Cong, Polyhedral-based data reuse optimization for configurable computing, in *Proceedings of the 21st ACM/SIGDA International Symposium on Field Programmable Gate Arrays* (Monterey, CA, 2013), pp. 29–38
10. W. Zuo, Y. Liang, P. Li, K. Rupnow, D. Chen, J. Cong, Improving high level synthesis optimization opportunity through polyhedral transformations, in *Proceedings of the 21st ACM/SIGDA International Symposium on Field Programmable Gate Arrays* (Monterey, CA, 2013), pp. 9–18
11. L.P. Carloni, K.L. McMillan, A.L. Sangiovanni-Vincentelli, Theory of latency-insensitive design. IEEE Trans. Comput.-Aided Des. Integrated Circ. Syst. **20**(9), 1059–1576 (2001)
12. J. Cortadella, M. Kishinevsky, B. Grundmann, Synthesis of synchronous elastic architectures, in *Proceedings of the 43rd Design Automation Conference* (San Francisco, CA, 2006), pp. 657–662
13. S.A. Edwards, R. Townsend, M.A. Kim, Compositional dataflow circuits, in *Proceedings of the 15th ACM-IEEE International Conference on Formal Methods and Models for System Design* (Vienna, 2017), pp. 175–184
14. R. Townsend, M.A. Kim, S.A. Edwards, From functional programs to pipelined dataflow circuits, in *Proceedings of the 26th International Conference on Compiler Construction*, (Austin, TX, 2017), pp. 76–86.
15. M. Budiu, S.C. Goldstein, *Pegasus: An Efficient Intermediate Representation* (Carnegie Mellon University, Tech. Rep. CMU-CS-02-107, 2002)
16. L. Josipović, R. Ghosal, P. Ienne, Dynamically scheduled high-level synthesis, in *Proceedings of the 26th ACM/SIGDA International Symposium on Field Programmable Gate Arrays* (Monterey, CA, 2018), pp. 127–136.

17. L. Josipović, A. Guerrieri, P. Ienne, Dynamatic: From C/C++ to dynamically scheduled circuits, in *Proceedings of the 28th ACM/SIGDA International Symposium on Field Programmable Gate Arrays* (Seaside, CA, 2020), pp. 1–10
18. L. Josipović, A. Guerrieri, P. Ienne, Synthesizing general-purpose code into dynamically scheduled circuits. IEEE Circ. Syst. Magaz. **21**(1), 97–118 (2021)
19. L. Josipović, A. Guerrieri, P. Ienne, From C/C++ code to high-performance dataflow circuits. IEEE Trans. Comput.-Aided Des. Integrat. Circ. Syst. **41**(7), 2142–2155 (2022)
20. G. Venkataramani, S.C. Goldstein, Leveraging protocol knowledge in slack matching, in *Proceedings of the 25th International Conference on Computer-Aided Design* (San Jose, CA, 2006), pp. 724–729
21. M. Najibi, P.A. Beerel, Slack matching mode-based asynchronous circuits for average-case performance, in *Proceedings of the 32nd International Conference on Computer-Aided Design* (San Jose, CA, 2013), pp. 219–225
22. L. Josipović, S. Sheikhha, A. Guerrieri, P. Ienne, J. Cortadella, Buffer placement and sizing for high-performance dataflow circuits, in *Proceedings of the 28th ACM/SIGDA International Symposium on Field Programmable Gate Arrays* (Seaside, CA, 2020), pp. 186–196
23. C. Rizzi, A. Guerrieri, P. Ienne, L. Josipović, A comprehensive timing model for accurate frequency tuning in dataflow circuits, in *Proceedings of the 22nd International Conference on Field-Programmable Logic and Applications* (Belfast, UK, 2022), pp. 375–383
24. L. Josipović, A. Marmet, A. Guerrieri, P. Ienne, Resource sharing in dataflow circuits, in *Proceedings of the 30th IEEE Symposium on Field-Programmable Custom Computing Machines* (New York, 2022), pp. 1–9
25. L. Josipović, P. Brisk, P. Ienne, An out-of-order load-store queue for spatial computing. ACM Trans. Embedded Comput. Syst. **16**(5s), 125:1–125:19 (2017)
26. L. Josipović, A. Bhattacharyya, A. Guerrieri, P. Ienne, Shrink it or shed it! minimize the use of LSQs in dataflow designs, in *Proceedings of the IEEE International Conference on Field Programmable Technology* (Tianjin, 2019), pp. 197–205
27. J. Liu, C. Rizzi, L. Josipović, Load-store queue sizing for efficient dataflow circuits, in *Proceedings of the IEEE International Conference on Field Programmable Technology* (Hong Kong, 2022), pp. 1–9
28. L. Josipović, A. Guerrieri, P. Ienne, Speculative dataflow circuits, in *Proceedings of the 27th ACM/SIGDA International Symposium on Field Programmable Gate Arrays*, (Seaside, CA, Feb. 2019), pp. 162–71.
29. M.S. Lam, Software pipelining: an effective scheduling technique for VLIW machines, in *Proceedings of the 1988 ACM Conference on Programming Language Design and Implementation* (Atlanta, GA, 1988), pp. 318–328
30. J.L. Hennessy, D.A. Patterson, *Computer Architecture: A Quantitative Approach*, 5th ed. (Morgan Kaufmann, 2011)
31. *Vitis High-Level Synthesis User Guide*, (AMD, 2022). [Online]. Available: https://docs.xilinx.com/r/en-US/ug1399-vitis-hls
32. *Intel HLS Compiler Pro Edition Reference Manual*, (Intel, 2022). [Online]. Available: https://www.intel.com/content/www/us/en/docs/programmable/683349/22-3/pro-edition-reference-manual.html
33. A. Canis, J. Choi, M. Aldham, V. Zhang, A. Kammoona, T. Czajkowski, S.D. Brown, J.H. Anderson, LegUp: an open-source high-level synthesis tool for FPGA-based processor/accelerator systems,. ACM Trans Embedded Comput. Syst. **13**(2), 24:1–24:27 (2013)
34. J. Choi, S. Brown, J. Anderson, From software threads to parallel hardware in high-level synthesis for FPGAs, in *Proceedings of the IEEE International Conference on Field Programmable Technology* (Kyoto, 2013), pp. 270–277
35. F. Ferrandi, V.G. Castellana, S. Curzel, P. Fezzardi, M. Fiorito, M. Lattuada, M. Minutoli, C. Pilato, A. Tumeo, Bambu: an open-source research framework for the high-level synthesis of complex applications, in *Proceedings of the 58th Design Automation Conference* (Virtual, 2021), pp. 1327–1330

36. J. Cheng, L. Josipović, G.A. Constantinides, P. Ienne, J. Wickerson, Combining dynamic and static scheduling in high-level synthesis, in *Proceedings of the 28th ACM/SIGDA International Symposium on Field Programmable Gate Arrays* (Seaside, CA, 2020), pp. 288–298
37. *http://www.llvm.org*, The LLVM Compiler Infrastructure, (2018). [Online]. Available: http://www.llvm.org
38. *https://mlir.llvm.org/*, Multi-Level IR Compiler Framework, (2020). [Online]. Available: https://mlir.llvm.org/
39. *https://github.com/llvm/circt*, CIRCT IR Compiler and Tools, (2020). [Online]. Available: https://github.com/llvm/circt
40. R. Nigam, S. Thomas, Z. Li, A. Sampson, A compiler infrastructure for accelerator generators, in *Proceedings of the 26th ACM International Conference on Architectural Support for Programming Languages and Operating Systems* (Virtual, 2021), pp. 804–817
41. H. Ye, C. Hao, J. Cheng, H. Jeong, J. Huang, S. Neuendorffer, D. Chen, ScaleHLS: a new scalable high-level synthesis framework on multi-level intermediate representation, in *Proceedings of the IEEE International Symposium on High-Performance Computer Architecture* (Seoul, 2022), pp. 741–755
42. W.S. Moses, L. Chelini, R. Zhao, O. Zinenko, Polygeist: raising C to polyhedral MLIR, in *Proceedings of the ACM International Conference on Parallel Architectures and Compilation Techniques* (Virtual, 2021), pp. 45–59
43. N. George, H. Lee, D. Novo, T. Rompf, K. Brown, A. Sujeeth, M. Odersky, K. Olukotun, P. Ienne, Hardware system synthesis from domain-specific languages, in *Proceedings of the 23rd International Conference on Field-Programmable Logic and Applications* (Munich, 2014), pp. 1–8
44. J. Ragan-Kelley, C. Barnes, A. Adams, S. Paris, F. Durand, S. Amarasinghe, Halide: a language and compiler for optimizing parallelism, locality, and recomputation in image processing pipelines. ACM Sigplan Notices **48**(6), 519–530 (2013)
45. J. Li, Y. Chi, J. Cong, HeteroHalide: from image processing DSL to efficient FPGA acceleration, in *Proceedings of the 2020 ACM/SIGDA International Symposium on Field-Programmable Gate Arrays* (Seaside, CA, 2020), pp. 51–57.
46. J. Pu, S. Bell, X. Yang, J. Setter, S. Richardson, J. Ragan-Kelley, M. Horowitz, Programming heterogeneous systems from an image processing DSL. ACM Trans. Arch. Code Opt. **14**(3), 1–25 (2017)
47. Y.-H. Lai, Y. Chi, Y. Hu, J. Wang, C. H. Yu, Y. Zhou, J. Cong, Z. Zhang, HeteroCL: a multi-paradigm programming infrastructure for software-defined reconfigurable computing, in *Proceedings of the 27th ACM/SIGDA International Symposium on Field Programmable Gate Arrays* (Seaside, CA, 2019), pp. 242–251
48. D. Koeplinger, M. Feldman, R. Prabhakar, Y. Zhang, S. Hadjis, R. Fiszel, T. Zhao, L. Nardi, A. Pedram, C. Kozyrakis et al., Spatial: a language and compiler for application accelerators, in *Proceedings of the 39th ACM SIGPLAN Conference on Programming Language Design and Implementation* (Philadelphia, PA, 2018), pp. 296–311.
49. J. Cong, J. Lau, G. Liu, S. Neuendorffer, P. Pan, K. Vissers, Z. Zhang, FPGA HLS today: successes, challenges, and opportunities. ACM Trans. Reconfigurable Tech. Syst. **15**(4), 1–42 (2022)
50. Y.-H. Lai, E. Ustun, S. Xiang, Z. Fang, H. Rong, Z. Zhang, Programming and synthesis for software-defined FPGA acceleration: status and future prospects. ACM Trans. Reconf. Tech. Syst. **14**(4), 1–39 (2021)
51. J. Cong, M. Huang, P. Pan, Y. Wang, P. Zhang, Source-to-source optimization for HLS. in *FPGAs for Software Programmers* (Springer, 2016), pp. 137–163.
52. J. Lau, A. Sivaraman, Q. Zhang, M.A. Gulzar, J. Cong, M. Kim, HeteroRefactor: refactoring for heterogeneous computing with FPGA, in *2020 IEEE/ACM 42nd International Conference on Software Engineering*, (Seoul, 2020), pp. 493–505
53. B.C. Schafer, Z. Wang, High-level synthesis design space exploration: past, present, and future. IEEE Trans. Comput.-Aided Des. Int. Circ. Syst. **39**(10), 2628–2639 (2020)

54. A. Sohrabizadeh, C.H. Yu, M. Gao, J. Cong, AutoDSE: enabling software programmers to design efficient FPGA accelerators. ACM Trans. Des. Automat. Electron. Syste. **27**(4), 1–27 (2022)
55. *Vivado High-Level Synthesis*, (Xilinx Inc., 2018). [Online]. Available: http://www.xilinx.com/products/design-tools/vivado/integration/esl-design.html
56. J. Cong, B. Liu, S. Neuendorffer, J. Noguera, K. Vissers, Z. Zhang, High-level synthesis for FPGAs: from prototyping to deployment. IEEE Trans. Comput.-Aided Des. Int. Circ. Syst. **30**(4), 473–491 (2011)
57. Y. Herklotz, Z. Du, N. Ramanathan, J. Wickerson, An empirical study of the reliability of high-level synthesis tools, in *2021 IEEE 29th Annual International Symposium on Field-Programmable Custom Computing Machines* (2021), pp. 219–223
58. F. Faissole, G.A. Constantinides, D. Thomas, Formalizing loop-carried dependencies in Coq for high-level synthesis, in *2019 IEEE 27th Annual International Symposium on Field-Programmable Custom Computing Machines*, (2019), pp. 315–315
59. J. Cheng, S.T. Fleming, Y.T. Chen, J. Anderson, J. Wickerson, G.A. Constantinides, Efficient memory arbitration in high-level synthesis from multi-threaded code. IEEE Trans. Comput. **71**(4), 933–946 (2022)
60. M. Geilen, T. Basten, S. Stuijk, Minimising buffer requirements of synchronous dataflow graphs with model checking, in *Proceedings of the 42nd Design Automation Conference* (Anaheim, CA, 2005), pp. 819–824
61. J. Xu, E. Murphy, J. Cortadella, L. Josipović, Eliminating excessive dynamism of dataflow circuits using model checking, in *Proceedings of the 31st ACM/SIGDA International Symposium on Field Programmable Gate Arrays* (Monterey, CA, 2023), pp. 27–37
62. J. Xu, L. Josipović, Automatic inductive invariant generation for scalable dataflow circuit verification, in *Proceedings of the 42nd IEEE/ACM International Conference on Computer-Aided Design* (San Francisco, CA, 2023, to appear)
63. M. Tan, S. Dai, U. Gupta, Z. Zhang, Mapping-aware constrained scheduling for LUT-based FPGAs, in *Proceedings of the 23rd ACM/SIGDA International Symposium on Field Programmable Gate Arrays* (Monterey, CA, 2015), pp. 190–199
64. C. Rizzi, A. Guerrieri, L. Josipović, An iterative method for mapping-aware frequency regulation in dataflow circuits, in *Proceedings of the 60rd ACM/IEEE Design Automation Conference* (San Francisco, CA, 2023, to appear)
65. H. Wang, C. Rizzi, L. Josipović, MapBuf: Simultaneous technology mapping and buffer insertion for hls performance optimization, in *Proceedings of the 42nd IEEE/ACM International Conference on Computer-Aided Design* (San Francisco, CA, 2023, to appear)
66. L. Guo, Y. Chi, J. Wang, J. Lau, W. Qiao, E. Ustun, Z. Zhang, J. Cong, Autobridge: coupling coarse-grained floorplanning and pipelining for high-frequency HLS design on multi-die FPGAs, in *Proceedings of the 29th ACM/SIGDA International Symposium on Field Programmable Gate Arrays* (Virtual, 2021), pp. 81–92
67. J. Zhao, T. Liang, S. Sinha, W. Zhang, Machine learning based routing congestion prediction in FPGA high-level synthesis, in *Proceedings of the Design, Automation and Test in Europe Conference and Exhibition* (Florence, 2019), pp. 1130–1135

Chapter 9
Logic Synthesis

Abstract This chapter delves into the subject of logic synthesis within the FPGA design process. It involves the conversion of high-level hardware description language (HDL) code, such as Verilog or VHDL, into a lower-level gate-level netlist that is suitable for FPGA implementation. The chapter covers the essentials of Boolean logic, logic optimization, technology mapping to FPGAs, and AI in logic synthesis. Once the logic synthesis process is complete, the resulting gate-level netlist serves as input to the FPGA physical implementation tools.

9.1 Overview

FPGAs have emerged as a popular platform for digital system design, offering a flexible, and reconfigurable alternative to Application Specific Integrated Circuits (ASICs). FPGA logic synthesis is a critical step in transforming high-level designs, usually described in Hardware Description Languages (HDLs), into a technology-specific implementation that can be mapped onto FPGA resources. This process involves several optimization and mapping techniques aimed at improving the performance, power consumption, and resource utilization of the final design.

Specifically, FPGA logic synthesis flow, starting from technology-independent optimization and progressing to FPGA technology mapping, explores the intricacies of combinational optimization and sequential optimization, both crucial aspects of synthesizing high-quality designs for FPGAs. The first stage in the synthesis flow is technology-independent optimization, which aims to improve the design without considering the specific target FPGA technology. This step involves two primary sub-steps:

1. Combinational Optimization—This process focuses on optimizing the combinational logic of the design, aiming to minimize the number of gates, levels, or interconnections. Techniques used for combinational optimization include constant propagation, Boolean minimization, and factoring. Various algorithms, such as Quine-McCluskey and Espresso, can be employed to achieve the desired optimization;

K. Tu et al., *FPGA EDA*, https://doi.org/10.1007/978-981-99-7755-0_9

2. Sequential Optimization—The sequential optimization step deals with the optimization of the design's sequential elements, such as registers and memory elements. Techniques used for sequential optimization include retiming, state encoding, and register minimization. These optimizations help to reduce the number of sequential elements, clock cycles, or overall latency of the design.

After technology-independent optimization, the design is mapped onto the specific target FPGA technology. The mapping process involves mapping Boolean logic to the FPGA's Lookup Tables (LUTs) and other combinational resources, such as adders and multipliers. This process involves matching the design's logic functions to the available FPGA resources while minimizing resource utilization, interconnect delay, and power consumption. The optimized sequential elements are mapped to the FPGA's flip-flops, latches, and other sequential resources during this step. The remainder of the chapter will delve into the details of logic synthesis for FPGA design.

9.2 Fundamentals of Boolean Logic

Boolean logic is a branch of mathematics that deals with logic operations on binary variables. The history of Boolean logic can be traced back to the work of George Boole in the 19th century, and its application to digital circuit design was formalized by Claude Shannon in the 1930s. Boolean logic has been used extensively in the design and analysis of digital circuits and electronic devices.

The essentials of Boolean logic include Boolean algebra, Boolean logic representation, and the basic logic gates. Boolean algebra is a mathematical system that deals with the operations and rules of logic. It is based on two binary values, typically represented as 0 and 1, and uses logical operators such as AND, OR, and NOT to perform logical operations on these values. Boolean algebra provides a set of rules and laws for manipulating logical expressions, which can be used to simplify complex logic circuits. Boolean logic representation is a way of expressing logical operations using symbols and diagrams. This representation uses logic gates to implement Boolean algebraic operations. The basic logic gates include AND, OR, and NOT gates, which can be combined to create more complex logic circuits.

9.2.1 Boolean Algebra

Boolean algebra is a branch of algebraic logic that deals with mathematical operations and expressions involving binary variables. The basic laws of Boolean algebra include identity, complement, associative, distributive, and commutative laws. For example, give Boolean variables A, B, and C. The distributive laws are $A(B + C) = AB + AC$ and $A + BC = (A + B)(A + C)$.

The Shannon expansion (decomposition) is a technique based on the distributive law of Boolean algebra to simplify complex logic expressions. Given a Boolean function $F(X) = F(x_1, \ldots, x_n)$ and $x_i \in X$ is a Boolean variable, then the identity $F = x_i F_{x_i} + \bar{x}_i F_{\bar{x}_i}$ holds. Where $F_{x_i} = F(x_1, \ldots, x_i = 1, \ldots, x_n)$ and $F_{\bar{x}_i} = F(x_1, \ldots, x_i = 0, \ldots, x_n)$ are called positive and negative cofactors of F with respect to x_i.

There are several operations with cofactors.

- *Boolean difference*: The Boolean difference or Boolean derivative of function F with respect to Boolean variable x is defined as $\partial F/\partial x = F_x \oplus F_{\bar{x}}$, that computes the difference between two Boolean expressions;
- *Universal quantification*: The universal quantification of F with respect to Boolean variable x is defined as $\forall x F = F_x F_{\bar{x}}$, that assigns variable x to a Boolean expression and requires that the expression is true for all possible values of the variable;
- *Existential quantification*: The existential quantification of F with respect to Boolean variable x is defined as $\exists x F = F_x + F_{\bar{x}}$, that assigns variable x to a Boolean expression and requires that the expression is true for at least one possible value of the variable.

Boolean difference and quantification are important concepts in Boolean algebra that have many practical applications in logic design, computer programming, and other fields that rely on digital circuits and logical operations.

9.2.2 Functional Representation

Truth Tables

The native way to represent a Boolean function is through its truth table. A truth table is a table that lists all possible combinations of inputs for a Boolean function and the corresponding output values for each combination. It is a native way to represent a Boolean function. An example of truth table is shown in Table 9.1, which represents the Boolean expression of the majority-of-three function

$$F = x_1x_2 + x_1x_3 + x_2x_3 \tag{9.1}$$

The hexadecimal form is thus encoded as `0xe8`.

Algebraic Expressions

Algebraic expressions consisting of Boolean variables and logical operators to represent the behavior of a Boolean function. There are two common types of Boolean algebraic expressions are sum-of-products (SOPs) and product-of-sums (POSs).

The SOP form, also known as disjunctive normal form (DNF), represents a Boolean function as the sum (OR) of multiple product (AND) terms. It is widely used for two-level logic synthesis. Typically, the general form of SOP is $f = \phi_1 \ldots \phi_n$, where ϕ_i represents a product term. The well-known Quine-McCluskey exact algorithms and ESPRESSO tool can be used for SOP minimization.

Table 9.1 An example of truth table

x_1	x_2	x_3	F
0	0	0	0
0	0	1	0
0	1	0	0
0	1	1	1
1	0	0	0
1	0	1	1
1	1	0	1
1	1	1	1

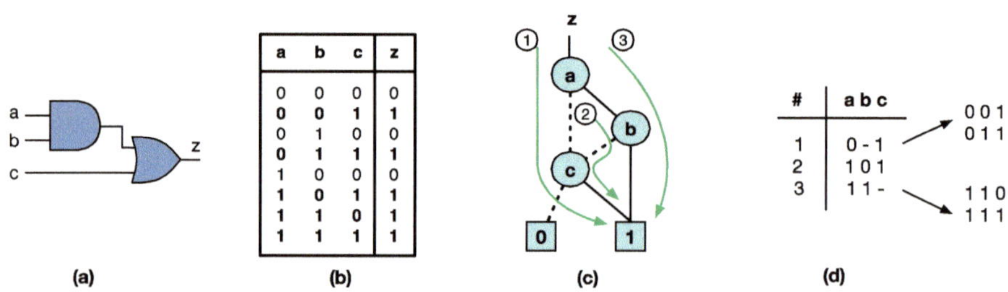

Fig. 9.1 *Boolean* function, *1-minterms*, and BDD. **a** Gate-level netlist, **b** corresponding truth table, **c** BDD of z, and **d** 1-minterms

In contrast, the POS form, also known as conjunction normal form (CNF), represents a Boolean function as the product (AND) of multiple sum (OR) terms. The general form of POS is $f = \psi_1 + \cdots + \psi_n$, where ψ_i represents a sum term. CNF is widely used as the input for SAT solving.

Both sum-of-products and product-of-sums forms are equivalent ways of representing Boolean functions, and one can be converted into the other using De Morgan's laws and other Boolean algebraic manipulations.

Binary Decision Diagrams

Bryant [1] introduced a concept of reduced, ordered BDDs (ROBDDs), along with a set of efficient operators for Boolean function manipulation (symbolic manipulation), and proved the canonicity property of ROBDDs (Fig. 9.1). BDDs are particularly useful because they allow for efficient manipulation of Boolean functions and are instrumental in optimization and verification tasks in digital circuit design. The fundamental building blocks of a BDD are its nodes and edges. Each internal node represents a Boolean variable, while the edges branching from the node signify the variable's value. Specifically, one branch (typically denoted by a solid line) stands for the variable being TRUE (or 1) and the other branch (often a dashed line) represents the variable being FALSE (or 0).

A minterm is a product term with all variables for which the function evaluates to 1. Minterms which produce output in 1(0) are called *1-minterms* (*0-minterms*). It is well know that finding all the 1-minterms (0-minterms) can be done by searching all the paths from the root to node 1(0) in the BDD (Fig. 9.1b) shows a truth table of the Boolean logic in Fig. 9.1a. This function has been converted into BDD, shown in Fig. 9.1c. The 1-minterms can be obtained by the following paths: (1) $\bar{a} \rightarrow$c; (2) a $\rightarrow \bar{b} \rightarrow c$; (3) $a \rightarrow b$. The 1-minterms are shown in Fig. 9.1d. We can see that these minterms may contain *don't cares* depending on the number of variables in each path.

BDD's unique graphical structure not only encapsulates the essence of Boolean decision-making but also paves the way for the optimization of digital circuits. Specifically, their role in logic minimization, functional decomposition, and technology mapping –optimizing and translating high-level descriptions to optimized and technology dependent implementations– are of paramount significance. Moreover, the canonical nature of ROBDD expedites equivalence checking, ensuring that circuit transformations throughout the design flow retain the intended functionality. However, while the BDD's utility in logic synthesis and EDA tools is indisputable, its efficacy can be curtailed by challenges such as potential size explosion and the intricacies of variable ordering. As synthesis tasks and EDA challenges evolve in complexity, striking a balance between the benefits offered by BDDs and the inherent challenges they present becomes critical, prompting researchers to continually refine methodologies and explore complementary representations.

9.2.3 Directed-Acyclic-Graph (DAGs) Boolean Network

A Boolean network is a directed acyclic graph (DAG) denoted as $G = (V, E)$ with nodes V representing logic gates (Boolean functions) and edges E representing the wire connection between gates. The input of a node is called its *fanin*, and the output of the node is called its *fanout*. The node $n \in V$ without incoming edges, i.e., no *fanins*, is the *primary input* (PI) to the graph, and the nodes without outgoing edges, i.e., no *fanouts*, are *primary outputs* (POs) to the graph. The nodes with incoming edges implement Boolean functions. The level of a node n is defined by the number of nodes on the longest structural path from any PI to the node inclusively, and the level of a node n is noted as $level(n)$ in this book.

9.2.3.1 And-Inverter Graph

And-Inverter Graph (AIG) [2] is one of the typical types of DAGs used for logic manipulation, where the nodes in AIGs are all two-inputs AND gates, and the edges represent whether the inverters are implemented. An arbitrary Boolean network can be transformed into an AIG by factoring the SOPs of the nodes, and the AND gates and OR gates in SOPs are converted to two-inputs AND gates and inverters with DeMorgan's rule. There are two primary metrics for evaluation of an AIG, i.e., *size*, which is the number of nodes (AND gates) in the graph, and *depth*, which is the number of nodes on the longest path from PI to PO (the largest level) in the graph. AIGs has also been extended for datapath synthesis and optimizations [3, 4].

A *cut* C of node n includes a set of nodes of the network. The nodes included in the *cut* of node n are called *leaves*, such that each path from a PI to node n passes through at least one leaf. The node n is called the *root* of the *cut* C. The cut size is the number of its leaves and the node itself. A cut is K-feasible if the number of nodes in the cut does not exceed K.

9.2.3.2 Majority-Inverter Graph

Majority-Inverter Graph (MIG) analogous to AIG [5]. The key difference lies in that the nodes in MIGs are all three-input majority (MAJ) gates, whose function is represented in (9.1). Let 'M' indicate the MAJ operation, i.e., $F = M(x_1, x_2, x_3)$. By setting any one of the variables to constants, the MAJ operation is reduced to AND/OR. For example, $F = M(x_1, x_2, 0) = x_1x_2$ and $F = M(x_1, x_2, 1) = x_1 + x_2$. Hence, AIGs can be converted to MIGs by one-to-one mapping.

For manipulation and analysis, MIG Boolean algebra is defined by including five transformation rules, referred to as Ω, to form an axiomatic system. Three rules are derived from Ω for the purpose of logic optimization, which is referred to as Ψ. The symbol $z_{x/y}$ represents a replacement operation by replacing x with y in all its appearance in z. A sound and complete axiomatization of majority-n (arbitrary odd number) logic is addressed in [6].

Despite the MIG, the M_5-inverter graph (M_5IG) is proposed in [7]. The experimental results show that M_5IGs obtain 10.4% improvement on size and 11.4% on depth compared to the method based on MIG. Although both M_5IGs and MIGs can be converted from AIG by one-to-one mapping, there are many constant inputs to the MAJ gate which cannot fully utilize the expression power of the MAJ gate. Exact synthesis can compute optimal MAJ expressions for small-scale logic functions. Combined with LUT-based logic resynthesis, the conversion based on a pre-computed optimal library produces better initial M_5IGs and MIGs than one-to-one mapping.

$$
\Omega \begin{cases}
\textbf{Commutativity}: \Omega.\text{C} \\
M(x, y, z) = M(y, x, z) = M(z, y, x) \\
\textbf{Majority}: \Omega.\text{M} \\
\begin{cases} \text{if}(x = y): M(x, x, z) = M(y, y, z) = x = y \\ \text{if}(x = y'): M\left(x, x', z\right) = z \end{cases} \\
\textbf{Associativity}: \Omega.\text{A} \\
M(x, u, M(y, u, z)) = M(z, u, M(y, u, x)) \\
\textbf{Distributivity}: \Omega.\text{D} \\
M(x, y, M(u, v, z)) = M(M(x, y, u), M(x, y, v), z) \\
\textbf{Inverter Propagation}: \Omega.\text{I} \\
M'(x, y, z) = M\left(x', y', z'\right)
\end{cases}
$$

$$
\Psi \begin{cases} \textbf{Relevance} - \Psi.\text{R} \\ M(x, y, z) = M(y, x, z_{x/y'}) \\ \textbf{Complementary Associativity} - \Psi.\text{C} \\ M(x, u, M(y, u', z)) = M(x, u, M(y, x, z)) \\ \textbf{Substitution} - \Psi.\text{S} \\ M(x, y, z) = \\ M(v, M(v', M_{v/u}(x, y, z), u), M(v', M_{v/u'}(x, y, z), u')) \end{cases}
$$

9.2.3.3 XOR-Majority Graph

The *XOR-Majority Graph* (XMG) is a modification of the MIG that incorporates XOR operations in addition to the existing MAJ gates and inverters [8]. By including two-input XOR gates, XMGs become more compact and are better suited for exact synthesis, which is sensitive to the number of nodes in the logic network. To optimize logic, several logic identities that explore MAJ-XOR Boolean equations are discussed in [9]. To further enhance the flexibility of XMG, the two-input XORs are expanded to three-input XORs, resulting in a generalized XMG that features both XOR and MAJ gates with three inputs [10].

9.3 Logic Optimization

9.3.1 Functional Methodologies

Truth Tables Karnaugh map method involves graphically representing the truth table using a Karnaugh map, which is a grid that allows for the grouping of adjacent cells that have the same output value. The grouped cells can be used to create a simplified Boolean expression, which can then be used to generate a minimized truth table. It is particularly effective for functions with fewer than six inputs, but can also be used for larger functions with some modifications.

Given a four-variable Boolean function $F(A, B, C, D) = \sum(1, 3, 4, 5, 7, 9, 11, 12, 14, 15)$. The Karnaugh map is shown in Fig. 9.2. By combining the group of minterms (1,3,9,11), (3,7,11,15), (4,5), and (12,14), the optimized Boolean expression is $F = \bar{B}D + CD + \bar{A}B\bar{C} + AB\bar{D}$.

Algebraic Expressions SOP expressions can be simplified using Boolean algebra rules, such as the distributive law, commutative law, associative law, and De Morgan's law. These rules can help to reduce the number of terms and/or variables used in the expression.

The Quine-McCluskey algorithm is a method for finding the minimum SOP expression for a given Boolean function, which is a computation version of Karnaugh map. The algorithm involves creating a table of all possible combinations of

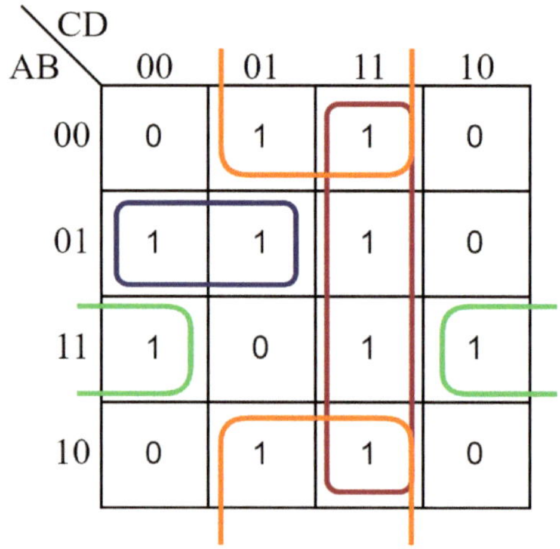

Fig. 9.2 Illustrative example of Karnaugh map optimization

input variables and grouping terms that differ by only one variable. By repeatedly applying the algorithm, it is possible to find the minimal SOP expression.

The ESPRESSO algorithm is a more efficient method for minimizing SOP expressions than the Quine-McCluskey algorithm. It is a heuristic algorithm that uses a combination of simplification rules and heuristics to find the minimal SOP expression. The algorithm has been shown to be more efficient than other methods for large complex functions.

9.3.2 DAG-Aware Logic Optimization

DAG-aware logic synthesis is an innovative approach to optimizing digital circuit designs, offering a unique perspective on the representation and manipulation of combinational and sequential elements. This technique leverages the power of directed-cycle graphs, a mathematical model that allows for the efficient encoding of complex circuit structures, to facilitate the synthesis process. Once the circuit is represented as a directed-cycle graph, various graph transformation techniques can be applied to optimize the design. These transformations include node merging, edge elimination, and subgraph replacement, among others. The objective of these transformations is to minimize the number of nodes and edges in the graph, effectively reducing the complexity of the circuit.

Rewriting, noted as `rw`, is a fast greedy algorithm for optimizing the graph size. It iteratively selects the AIG subgraph with the current node as the root node and replaces the selected subgraph with the same functional pre-computed subgraph with a smaller size to realize the graph size optimization. Specifically, it finds the 4-feasible cuts as subgraphs for the node and preserves the number of logic levels [2]. For example, Fig. 9.3 shows the optimization of the original graph in Fig. 9.3 with `rw`. The

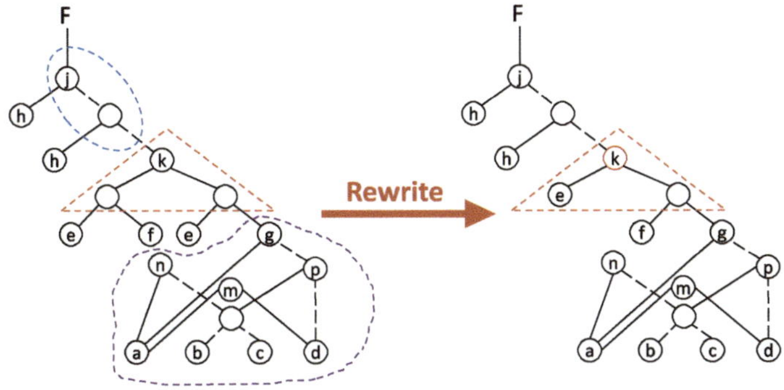

Fig. 9.3 And-inverter-graph logic rewriting

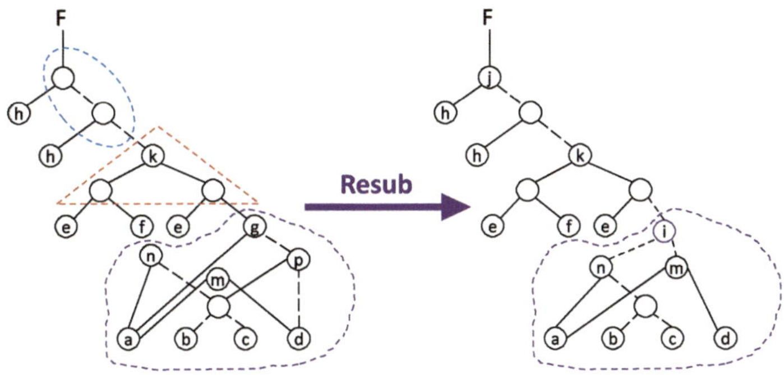

Fig. 9.4 And-inverter-graph logic refactoring

algorithm visits each node in the topological order and checks the transformability of its cut w.r.t `rw`. It will skip the node and visit the next one if not applicable. In Fig. 9.3, the node j is skipped for `rw`, and node $k = efr$ is optimized with `rw`, resulting in the node reduction of 2 for the AIG optimization.

Refactoring, noted as `rf`, is a variation of the AIG `rewriting` using a heuristic algorithm [11] to produce a large cut for each AIG node. Refactoring optimizes AIGs by replacing the current AIG structure with a factored form of the cut function. It can also optimize the AIGs with the graph depth. For example, Fig. 9.4 shows the optimization of the original AIG with `rf`. The node $j = h(\overline{h} + k)$ is optimized with `rf` based on DeMorgan's rule, where $j = hk$. As a result, the optimized graph with `rf` has a graph size of 19 with 2 nodes reduction and 1 depth reduction.

Resubstitution, noted as `rs`, optimizes the AIG by replacing the function of the node with the other existing nodes (`divisors`) already present in the graph, which is expected to remove the redundant node in expressing the function of the current node. For example, in Fig. 9.5, the node $g = a\overline{p}$, $p = \overline{b}\,\overline{c}\,\overline{d}$, i.e., $g = a(b + c + d)$, together with existing divisors $m = ad$, $n = a(b + c)$, the node g can be resubstituted with

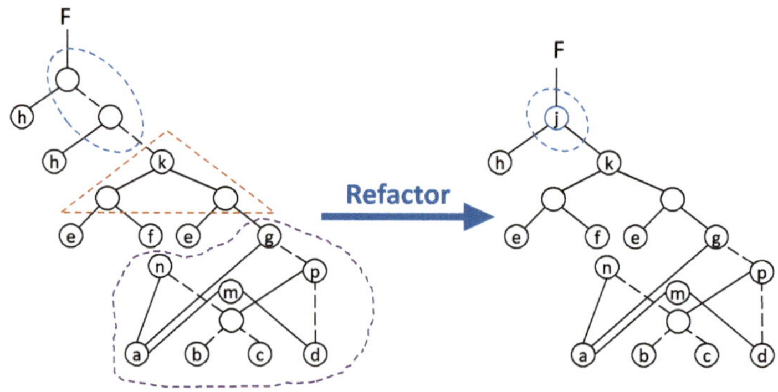

Fig. 9.5 And-inverter-graph logic resubstitution

the complement of node i, where $\overline{i} = m + n = a(b + c + d)$, and node p is removed from the graph. As a result, with `rs`, the original AIG is optimized in graph size by 1 node reduction.

9.3.3 Exact Logic Optimization

Compared to heuristic algorithm-based logic synthesis, research on exact synthesis is much less with the most difficult problem being the high computational complexity, which requires enumeration to complete decision and reasoning. Due to computational complexity, the heuristic optimization method Espresso is more popular than the Quine-McCluskey algorithm, a well-known two-level logic sum-of-products (SOP) minimization algorithm [12]. Exact synthesis requires a given Boolean function ($\mathcal{F}$), a set of logic gates or primitives used to represent the function ($\mathcal{L}$), and a cost function ($\mathcal{C}$).

From the perspective of the methods used for exact synthesis, the traditional exact synthesis includes the following three methods [13].

1. An arbitrary Boolean function can be decomposed into logic expressions by functions. A typical example is Shannon decomposition. The function can be decomposed as $f = xF_x + \bar{x}F_{\bar{x}}$, where $F_x = f(x = 1)$ and $F_{\bar{x}} = f(x = 0)$ are the positive and negative cofactors of the f. The logic expression can be obtained by recursively calling Shannon decomposition on the logic function. In addition, there are other decomposition methods such as Disjoint-Support Decomposition (DSD) and bi-decomposition. Through different decomposition methods to obtain the logical expression from which to optimize the optimal solution.
2. Given the set of logic operation symbols $\mathcal{L}$, the number of logic networks (n) composed of logic gates is limited. Through enumerating the logic network and assigning operators to the logic network nodes, different Boolean functions can be obtained, and the obtained Boolean functions are compared with the synthesized

Boolean functions. If they are consistent, the optimal solution is found, otherwise, the n is increased in the next enumeration. The n starts from 0, it is clear that the number of enumerated logic networks grows non-linearly with increasing.

3. The method 1 and the method 2 are combined to perform function decomposition and logic network enumeration simultaneously.

However, these methods can only deal with a very limited number of Boolean functions and can only prove optimality by enumeration. In recent years, with the improvement of computing power and the development of theoretical computer science, the exact synthesis methods of decision and reasoning using constraint solvers, especially Boolean satisfiability (SAT) solvers, have attracted much attention. SAT-based exact synthesis was proposed at the 2007 Formal Methods in Computer-Aided Design (FMCAD) conference [14]. Subsequently, Knuth et al. used different CNF codes for exact synthesis, but only limited to 2-input operator optimal Boolean chain, and these algorithms all aim to find the Boolean chain with optimal area (minimum number of Boolean chain links) [15]. Mathias et al. extended them to synthesize Boolean chains with optimal logic depth, using a SAT solver to check whether there exists a Boolean logic network with the fewest logic levels that implements a given function under the constraint [16]. The task of SAT-based exact synthesis is therefore to find the optimal logical expression for a given input constraint in terms of size or depth.

9.3.4 Exact Synthesis Algorithm Flow

Figure 9.6 shows the flow of the exact synthesis algorithm for solving the optimal area. The core idea is to call the solver to answer the question "Is there a set of Boolean functions $\mathcal{F}$ to be synthesized by a logic network composed of r logic gates belonging to the set $\mathcal{L}$, where $C = r$?" by constraint coding. Therefore, the initial value of r is 0, which is mainly considering that some functions to be synthesized are constants 0 or 1, primary input or its inverse variables. No logic gate is needed at this time, such as $\mathcal{F} = 1$, x_1 is the primary input, and the function to be synthesized is a simple case that does not need a logic gate. Under the given constraint r, the CNF-based SAT coding $\mathcal{F}_r$ is generated and delivered to the SAT solver for solving. If the SAT solver returns a satisfiable solution, the optimal logic network for realizing the function to be synthesized is found; otherwise, r is increased by 1, and the $\mathcal{F}_r'$ will guide SAT to return the satisfiable solution again. Therefore, the solution can be ensured to be optimal, that is, the logic function to be synthesize is realized by using the least logic gates. A given upper bound of r or a time constraint on the SAT solver can cause the algorithm to terminate prematurely.

From the flow of the algorithm, it is the most important step to encode SAT under the constraints of r and logic gate set $\mathcal{L}$. In addition, the complexity of the algorithm is closely related to the size of r. When r is small, the number of variables and clauses of SAT coding is relatively small, so SAT solver can often return results quickly. If a function is very complex and requires more logic gates to implement, it

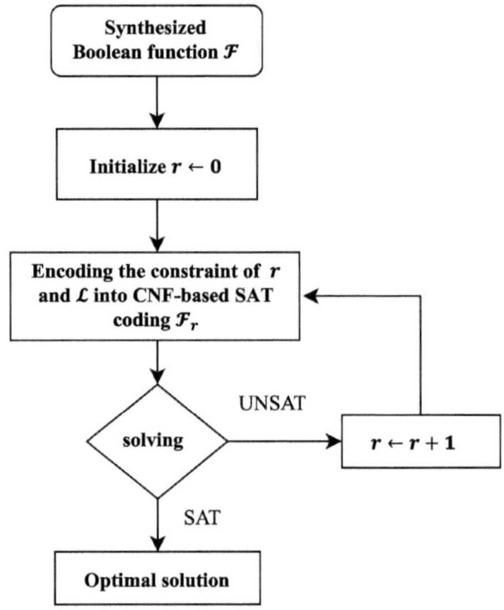

Fig. 9.6 Bottom-up flow of SAT-based exact synthesis

is obvious that the difficulty of solving the SAT solver will increase with the number of SAT encoding variables and the number of clauses. For example, assuming that the optimal solution is $r = 10$, then the SAT is trying to prove that these are unsatisfied (UNSAT) in the bottom-up increment process of $r = 9$.

Compared with the invalid search caused by bottom-up increase r, the [17] proposed the synthesis process of top-down decrease r as shown in Fig. 9.7. Firstly, the upper bound of the number of nodes r of the function to be synthesized is obtained through function decomposition, and the SAT solver is called to answer the question "Is there a set of Boolean functions $\mathcal{F}$ to be synthesized by a logic network of $r = r - 1$ logic gates belonging to the set $\mathcal{L}$?". If it is UNSAT, then the $r + 1$ is already the optimal value; otherwise, it continues to decrease r until the SAT solver returns UNSAT. In the same way, how to obtain a more compact upper bound r is crucial.

9.3.5 SAT-Based Encoding

The two core tasks of SAT coding are defining Boolean variables and adding constraints based on these defined Boolean variables. The Boolean function $f(x_1, x_2, \ldots, x_n) = f(0, 0, \ldots, 0) = 0$ is defined as a normal function. If all the steps in a Boolean chain are normal functions, the Boolean chain is called a normal Boolean chain. For example, if a Boolean chain consists of "AND/OR", it is a normal Boolean chain; a Boolean chain is not normal if it contains a NAND/NOR. An abnormal Boolean function can be made the inverse of a normal function by adding

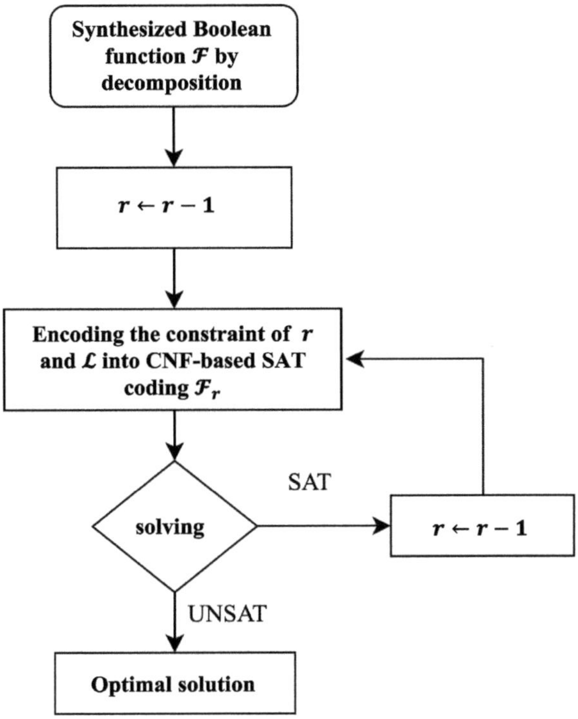

Fig. 9.7 Top-down flow of SAT-based exact synthesis

an inverter, such as $f(a, b) = \overline{a + b}$, where f is an abnormal function, but $\bar{f}$ is a normal function. In this subsection, the CNF coding proposed by Knuth [15] for 2-input conventional Boolean chain is taken as an example to illustrate.

9.3.5.1 Variable Definition

If the primary input of the logic function to be synthesized is n, the primary output number is m, and the current number of logic gates to be encoded is r, for $1 \leq h \leq m$, $n < i \leq n + r$ and $0 < t < 2^t$, respectively define the following Boolean variables:

(i) x_{it}, which represents the tth bit of the logic gate truth table of x_i.
(ii) g_{hi}, denotes that the logic gate x_i is the hth output of the Boolean function.
(iii) s_{ijk}, denotes that the inputs to the logic gate x_i are x_j and x_k, where $1 \leq j < k < i$.
(iv) f_{ipq}, represents the output Boolean value of the logic gate x_i when it is (p, q). To reduce the number of variables, the f_{i00} in f_{ipq} does not appear because $f_i(00) = 0$ in a normal Boolean chain.

9.3.5.2 Constraints

Constraints include basic constraints and extended constraints, in which the basic constraints ensure that the correct results can be obtained, while the extended constraints add symmetry breaking rules to reduce the search space through the characteristics of the logical network. Due to space constraints, this paper only introduces the basic constraints.

The core of the basic constraints is to ensure that the logic gates in the Boolean chain complete the correct logic operation, which is also called the main constraint clause. For $0 \leq a, b, c \leq 1$ and $1 \leq j < k < i$, the primary-subclause constraint is defined as

$$(s_{ijk} \wedge (x_{it} \oplus \bar{a}) \wedge (x_{jt} \oplus \bar{b}) \wedge (x_{kt} \oplus \bar{c})) \rightarrow (f_{ibc} \oplus \bar{a}),$$

where it is the implication logic expression, that is, $f = y_1 \rightarrow y_2 = \bar{y_1} + y_2$. It can be transformed into CNF form as

$$(\bar{s}_{ijk} \vee (x_{it} \oplus a) \vee (x_{jt} \oplus b) \vee (x_{kt} \oplus c)) \vee (f_{ibc} \oplus \bar{a}),$$

where this constraint can be translated as, if the inputs of x_i are x_j and x_k, and the values of x_i, x_j, and x_k in the tth truth table are a, b, and c, then the x_i must perform the logical operation $b \circ_i c = a$. The a, b, and c are constants that control the polarity of the variable.

Furthermore, if the logic gate x_i is the output, then the truth table of x_i must be equal to the Boolean function in which it is to be synthesized, thus adding a constraint as

$$\bar{g}_{hi} \vee (x_{it} \oplus g_h(t_1, t_2, \ldots, t_n)),$$

where $(t_1, t_2, \ldots, t_n)_2$ is the binary encoding of t. For each output function, add constraints $\vee_{i=n+1}^{n+r} g_{hi}$ to ensure that one step is output. Finally, constraints $\vee_{k=1}^{i-1} \vee_{j=1}^{k-1} s_{ijk}$ are added to ensure that each step in the Boolean chain has two valid inputs.

9.3.5.3 Example

Take the Boolean chain shown in the Fig. 9.8 as an example, the variable x_i encodes the global function of each logic gate according to the truth table. Since $n = 3$, $m = 2$, $r = 2$, the number of the logic gate belongs to the interval $(n, n + r]$, i.e., x_4 and x_5, with a total of $2^n - 1 = 7$ truth table bits. Note that the 0 bit does not need to be encoded because of the conventional function.

$$\begin{array}{lcccccccc} \text{t} & = & 7 & 6 & 5 & 4 & 3 & 2 & 1 \\ x_{4t} & = & 1 & 0 & 0 & 0 & 1 & 0 & 0 \\ x_{5t} & = & 0 & 1 & 1 & 1 & 1 & 0 & 0 \end{array}$$

Fig. 9.8 Example of Boolean chain

There are four variables g_{hi}, two of which are assigned 1, indicating the corresponding logic gate of the function input, so that $g_{14} = 1$, $g_{15} = 0$, $g_{24} = 0$, $g_{25} = 1$. There are 9 selection variables, assigned as

k	=	2	3	4
s_{41k}	=	1	0	
s_{42k}	=		0	
s_{51k}	=	0	0	0
s_{52k}	=		0	0
s_{53k}	=			1

Finally, the variable f_{ipq} encodes the truth table of the AND or XOR gate as

(p,q)	=	(1,1)	(0,1)	(1,0)
f_{4pq}	=	1	0	0
f_{5pq}	=	0	1	1

9.3.6 Sequential Logic Optimization

Retiming is a sequential optimization technique that has been studied since 1980's. Retiming techniques optimize the sequential circuits by relocating *edge-triggered registers*[1] across the combination logic without changing the design functionality. A lot of research efforts have been spent on developing promising retiming techniques that mainly target on three objectives:

- *min-delay*: minimize the worst-path delay of the circuits [18];
- *min-area*: minimize the number of registers of the circuits [19];
- *constrained min-area*: minimize the number of registers with a given worst-path delay constrain [20].

Numerous techniques have been proposed in our community to achieve these three objectives [18, 19, 21–24], and have been demonstrated with encouraging results. Although retiming assumes that the topology of the combinational logic is fixed, the quality (at logic-level) of the design can be further improved by combining the combinational logic optimization techniques and technology mapping [25][26][27]. In practice, constrained min-area retiming, has been incorporated into end-to-end design flows, which targets on improving the performance of the design regarding the delay, power, area, etc.

In 1997, N. Shenoy published the first retiming survey [28]. This work reviewed the theories and practical implementations of retiming, and the side issues of incorporating in the design flow. Due to the significant changes in the technology and design complexity, an up-to-date industrial study of retiming is necessary. Moreover, to our knowledge, no credible work ever evaluates the retiming algorithms in an end-to-end design flow. Most existing retiming algorithms were evaluated at the end of the logic synthesis, where the delay and area are measured after technology mapping or using unit delay-area models. However, due to the significant increase in design complexity and design rules, the gate-level netlist does not correlate well with the physical netlist [4]. Hence, in this work, the experimental results are collected at the end of the physical design process. Note that the negative retiming operations[2] are forbidden in our experiments. It turns out that the performance improvements gained by retiming evaluated the logic level could make the final physical netlist worse. Also, there are significant extra design efforts required for retimed designs, e.g., sequential equivalence checking (SEC). Thus, for the designs that retiming does not provide enough improvements in design performance, retiming needs to be avoided in the design flow. These give us the motivation for developing a *prediction* mechanism for retiming in real-world design flow, e.g., using machine learning techniques.

The concept and the three objectives of retiming are illustrated using a simple example shown in Fig. 9.9. The original design is shown in Fig. 9.9(1). We assume all the gates have unit delay one, and the edge-triggered registers are represented using the rectangles. The original design has five registers ($n = 5$) and the delay of the critical path, {a,b} $\rightarrow f$, is four ($d = 4$).

The *min-delay* retiming result is shown in Fig. 9.9(2). There are two iterations in this retiming: (1) move the two registers connected with a and b forward, which

[1] In the rest of this paper, *register* is used to represent *edge-triggered registers*.

[2] For min-delay retiming, negative retiming refers to "the delay of critical path increases"; for min-area retiming, it refers to "the number of registers increases."

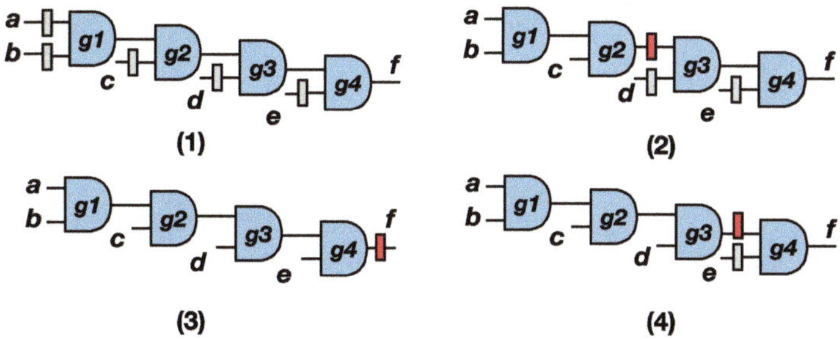

Fig. 9.9 Illustrative examples of retiming: (1) original netlist; (2) *min-delay* retiming; (3) *min-area* retiming; (4) *min-area* retiming under delay constrain (delay$\leq$3)

makes gate g_2 retimable; (2) move the retimed register and the register connected with c forward. The delay of the retimed design in Fig. 9.9(2) is two ($d = 2$), and the number of registers is three ($n = 3$). The *min-area* retiming result is shown in Fig. 9.9(3). This solution requires two more iterations in addition to the solution of min-delay retiming, which move the registers all the way to primary output and reduce the number of registers to one. The delay is increased to four. The third objective is min-area retiming under delay constrain. In this example, let the delay constraint be $d \leq 3$. The solution is shown in Fig. 9.9(4). We can see that the min-delay retiming gives the best delay solution ($d = 2$), min-area offers the minimum number of registers, and min-area retiming with delay constraints gives balanced results in between of min-area and min-delay.

9.3.6.1 Formulation of Retiming

Most logic optimization techniques are formulated based on direct graph representation. The logic netlist, so called *Boolean network* can be modeled using direct graph $G(V, E)$, where each vertex v corresponds to a logic gate g in the design. Besides constructing the direct graph directly from the netlist, this can also be done based on the transformed Boolean network, such as *And-Inv-Graph* [29] and its sequential version [30]. In the context of retiming, the sequential Boolean network is the combinational network separated by the memory elements, which are assumed to be ideal registers. The edges in the graph $G(V, E)$ represent the interconnections of the logic gates in the design.

Let us denote e_{uv} is an edge of $G(V, E)$, e_{uv}: $u \rightarrow v$, and w_{uv} is the weight of the edge e_{uv} which represents the number of registers between the two vertex u and v. The weight of the edges directed from and into the primary inputs (PIs) and primary outputs (POs) is zero. Each vertex v in $G(V, E)$ represents its delay of the corresponding gate, denoted $d(v)$. The problem of retiming is denoted by retiming lag function $r(v)$ [31]: $V \rightarrow Z$. Let us denote that w^r_{uv} is the weight of the edge e_{uv}. For any retiming, it can be represented by Eq. 9.2.

$$w^r_{uv} = r(v) - r(u) + w_{uv} \tag{9.2}$$

The value of $r(v)$ represents the movement of the registers for vertex v. If it is forward retiming (from inputs to outputs), then $r(v)$ is a negative. For any legal retiming, the condition shown in Eq. 9.3 must be satisfiable.

$$w_{uv} + r(v) - r(u) \geq 0 \tag{9.3}$$

9.3.6.2 Min-Delay Retiming

The min-delay retiming problem is as follows: Given $G(V, E)$ with a vertex delay function d and edge weight function w, find a legal retiming r, such that the cycle time c is minimized:

$$c = \max_{p:w_r(p)=0} \{d(p)\}, \tag{9.4}$$

where $d(p)$ is the path delay, and $w_r(p)$ is the retimed register count on the path p. To this problem, Leiserson and Saxe [31] developed a classic algorithm: Two matrices W and D are first defined as

$$W(u, v) = \min_{p:u \rightsquigarrow v} \{w(p)\}, \tag{9.5}$$

$$D(u, v) = \max_{p:u \rightsquigarrow v \wedge w(p)=W(u,v)} \{d(p)\}, \tag{9.6}$$

$W(u, v)$ gives the minimum register count on any path from u to v. $D(u, v)$ determines the maximum delay from u to v for the minimum register count. The two matrices can be obtained by solving an all-pairs shortest paths problem in G. Afterward, a binary search for the minimum clock cycle is performed. In each iteration, a Bellman-Ford algorithm can be employed to test whether a legal retiming exists with the current cycle time c. The algorithm above runs in $O(V^3 \lg V)$ time because each iteration costs $O(V^3)$ time for a Bellman-Ford algorithm and the binary search runs in $O(\lg V)$.

Leiserson and Saxe [31] also proposed another more efficient relaxation algorithm, which runs in $O(VE)$ for examining if a retiming exist for a given clock cycle c. A function $\Delta(v)$ gives the largest delay seen from any path that terminates at the output of v:

$$\Delta(v) = d(v) + \max_{u \in FI(v),\ w(e_{uv})=0} \{\Delta(u)\}. \tag{9.7}$$

Therefore, the cycle time can be expressed as follows:

$$c = \max_{v \in V} \{\Delta(v)\}. \tag{9.8}$$

The relaxation algorithm consist of alternately updating the functions $\Delta(v)$ and $r(v)$ for $|V| - 1$ times. The optimality is guaranteed because each iteration simulates a pass off a Bellman-Ford algorithm (i.e., a vertex being relaxed in a pass of the Bellman-Ford algorithm must be updated in an iteration of the relaxation algorithm), one can obtain a feasible retiming under the target cycle time c, if it exists. Because calculating $\Delta(v)$ in an iteration costs $O(E)$ time, the relaxation algorithm runs in $O(VE)$ time. Later on, the runtime of this algorithm was improved by Shenoy and Rudell [32] by adding an early break mechanism.

9.3.6.3 Min-Area Retiming

The typical min-area retiming refers to minimizing the number of registers without delay constraints. In which case, the problem can be formulated as a minimum cost flow problem using linear programming. The formulation is as follows:

$$\min : \sum_{\forall e_{uv}} r(u) - r(v) \wedge (\forall e_{uv}, r(u) - r(v) \leq w_{uv}) \tag{9.9}$$

Goldberg [23] presented a practical push-relabel method that can solve this problem in $O(V^2E\ \log(\text{VC}))$ worst-case runtime, where V is the number of vertices, E is the number of edges, and C is the maximum cost of the edges. A. Hurst et al. [19] proposed a the min-area retiming using maximum network flow problem. It turns out that within a combinational network, minimizing the number of registers by retiming is equivalent to finding a minimum cut. Note that the minimum cut problem is the dual of the maximum network flow problem. Computing maximum flow of a network is much less complex than minimum cost determination. Although there may exist many minimum cuts, that approach always generates one minimum cut that provides the minimum number of movements of the registers. This is claimed to simplify the computation of the initial states and minimize the side effects of the design [19]. The worst-case runtime of maximum flow approach is bounded by $O(R^2E)$, where R is the initial number of registers, and E is the number of edges. This algorithm requires repeated iterations while the number of iterations is typically small, as demonstrated by the authors.

9.3.6.4 Constrained Min-Area Retiming

Although it is claimed that pruning the redundant storage elements in the design reduces the area, power, and verification cost, most designs request a specific target clock period. The first attempt of constrained min-area retiming was presented by Shenoy et al. [32]. The implementation was composed by computing the W and D matrices, and the minimum cost circulation implementation. Let us denote $N_{fanin}(v)$=number of $fanins$ of vertex v, $N_{fanout}(v)$=number of $fanouts$ of vertex v. The formal definition of this problem is represented as follows [28]:

$$\min : \sum_{\forall v} |N_{fanin}(v) - N_{fanout}(v)| \cdot r(v) \tag{9.10}$$

$$\wedge \ \forall e_{uv}, r(u) - r(v) \leq w_{uv} \tag{9.11}$$

$$\wedge \ \forall e_{uv}, r(u) - r(v) \leq W(u, v) - 1 \tag{9.12}$$

Equation 9.9 represents the cost of number of registers of all register relocations. Equations 9.10 and 9.11 constrain each relocation must be legal retiming and under the delay constrain. We can see that this problem can be solved by combining the algorithms proposed in Sects. 9.3.6.3 and 9.3.6.2. The most recent constrained min-area retiming method was proposed by Hurst et al. [20], which the min-area approach is based on the work of [19]. That approach is developed based on the observation that area-critical and timing-critical regions are rare overlapped. In that work, the timing constrains of retiming requires the exiting min-delay retiming algorithms, which give the initial register positions for their min-area approach.

9.3.7 Advanced Logic Optimization Techniques

Logic Decomposition is a technique used in digital circuit design to break down a complex logic function into simpler sub-functions. The goal of logic decomposition is to simplify the design process and reduce the number of logic gates required to implement the function. The process of logic decomposition involves analyzing the function and identifying repeated sub-functions that can be extracted and implemented as separate modules. These sub-functions are typically implemented using smaller and simpler logic gates, such as AND, OR, and XOR gates.

Disjoint-support decomposition (DSD) is a powerful technique used to break down a Boolean function into a set of sub-functions that have non-overlapping sets of variables. The process of disjoint-support decomposition involves partitioning the variables of the Boolean function into disjoint sets. The function is then decomposed into a set of sub-functions, where each sub-function is defined over a distinct set of variables. The resulting sub-functions can be implemented as separate modules, which can be interconnected to form the original function. Despite the typical AND, OR, XOR, and 2-to-1 multiplexer (MUX) gates, the decomposition using majority-of-three gate is also addressed in [17, 33].

Boolean Matching is a technique used in logic synthesis to find common sub-expressions between the inputs and outputs of a digital circuit, and to simplify the circuit by eliminating redundant logic gates. It can be used to find and reutilize equivalent subcircuits in order to reduce the amount of work in each design iteration and accelerate design closure [34]. This is done by identifying common subcircuits in the design and replacing them with a single shared subcircuit. The process of identifying common subcircuits involves analyzing the Boolean expressions of the gates in the design and looking for patterns that can be combined or eliminated. This is typically done using automated software tools that can perform Boolean matching algorithms.

9.4 Technology Mapping

Technology mapping for FPGAs is the process of transforming a technology-independent logic network, called the subject graph, into a network of logic nodes, each of which can be realized as one K-input LUT. A traditional LUT can implement any Boolean function up to K inputs. The subject graph is often represented as an AND-Inverter Graph (AIG) composed of two-input ANDs and inverters. Most structural methods of FPGA mapping [35, 36] start by computing all cuts for each AIG node. Next, the AIG nodes are traversed in a topological order and a dynamic programming approach is used to find an optimum-depth LUT mapping of the AIG. This mapping can often be substantially improved by applying area-recovery heuristics [37–39] to reduce the number of LUTs while preserving the depth of the LUT network. It should be noted that such FPGA mapping algorithms as FlowMap [40] and CutMap [41] do not compute all cuts.

However, good cuts in these mappers are found using the maximum flow algorithm that has high-computational complexity. As a result recent state-of-the-art mappers use cut enumeration rather than maximum flow. In a large class of programmable architectures, the LUT size K varies between 3 and 6. For these relatively small LUT sizes, the traditional methods for LUT mapping based on cut enumeration work quite well. For K equal to 4 or 5, exhaustive cut enumeration can be applied, resulting in an average of 10–40 cuts stored at each node. When the LUT size is 6, exhaustive cut enumeration may lead to 100+ cuts per node. Cut representation takes substantial memory when mapping large Boolean networks. To remedy the situation, a partial cut enumeration can be used, which heuristically prune the cuts, resulting in reduced memory requirements. However, cut pruning may result in losing good cuts. In this case, the depth-optimality of mapping is not guaranteed.

Another class of modern programmable architectures realizes logic networks using macro-cells, which typically contains LUTs and other logic gates. A straightforward way of mapping logic into programmable macro-cells starts by computing all K-input cuts for each node where K is the number of macro-cell inputs. Unlike a K-input LUT, a macro-cell cannot implement all logic functions of K inputs. Therefore, the local function of each cut is computed in terms of the cut inputs, and only those cuts whose logic function can be expressed by the macro-cell will be kept for potential mapping. However, this approach is not practical because a macro-cell often has 10 or more inputs while the number of 10-input cuts is extremely large for all but the smallest benchmarks.

9.4.1 Flow-Based and Cut-Based LUT Mapping

Flow-based and cut-based LUT mapping are two approaches used in the process of logic optimization in digital circuit design.

In flow-based LUT mapping, the logic optimization is performed based on the flow of data through the circuit. The circuit is represented as a directed acyclic graph (DAG) where nodes represent combinational logic blocks and edges represent the data flow between the blocks. The optimization algorithm works by identifying paths in the DAG that can be merged or simplified to reduce the number of logic blocks and/or the delay of the circuit.

In cut-based LUT mapping, the logic optimization is performed based on the cuts in the circuit. A cut is a set of inputs to a logic block that, when fixed, determine the output of the block. The circuit is represented as a graph where nodes represent the logic blocks and edges represent the inputs and outputs of the blocks. The optimization algorithm works by identifying cuts in the graph and replacing them with LUTs that implement the same functionality.

Both flow-based and cut-based LUT mapping have their advantages and disadvantages. Flow-based LUT mapping is generally more effective in reducing the delay of the circuit but may not necessarily result in a reduction of the number of logic blocks. Cut-based LUT mapping, on the other hand, is generally more effective in reducing the number of logic blocks but may result in a larger delay. The choice between the two approaches depends on the specific requirements of the circuit being designed.

9.4.2 Cut-Less LUT Mapping

Cut-based technology mapping algorithms are known to be computationally demanding because they involve several steps such as cut enumeration, pruning, and computation of the local function of cuts and their canonical form. These steps are essential for the algorithm to work effectively, but they require significant computational resources, making the process time-consuming and expensive [42]. Cut-less LUT mapping was addressed in [43], in which only one cut is computed and stored at each AIG node, instead of computing all cuts. However, the resulting mapping may not necessarily be depth-optimal, unlike in traditional mapping methods. The experimental results show that with a runtime and memory complexity linear to the number of nodes in the subject graph, the cut-less LUT mapping algorithm performs effectively for LUTs with 12 or more inputs.

9.5 AI in Logic Synthesis

As previously mentioned, a major obstacle to swift hardware specialization is the absence of assurance provided by current FPGA CAD tools in achieving design closure without additional customization [44, 45]. The utilization of these tools typically necessitates considerable manual labor to fine-tune and adjust a vast array of design parameters and tool options to attain a high QoR. Regrettably, the evaluation of a single design point can be exceedingly time-consuming, as design stages such as place and route often take hours or even days to complete for large circuits. To facilitate agile FPGA-based compute acceleration, it is imperative to reduce design costs by:

1. substantially decreasing the time needed for accurate QoR estimation,
2. minimizing human intervention in the design tuning process.

In recent years, there has been a growing trend of employing machine learning (ML) techniques to expedite the FPGA design process and diminish the reliance on human engineering efforts. This approach is thought to hold significant promise in tackling more pressing challenges within FPGA design. From the perspective of contemporary FPGA design, the driving factors can be concisely outlined as follows:

- **Fast and accurate approximation via predictive modeling.** Machine learning can serve as a statistical technique that extracts domain knowledge from historical and existing data to predict future or unseen outcomes related to specific algorithmic or mathematical objectives. The recent advancements in machine learning algorithms and neural architectures enable the creation of generic and accurate approximations for given objectives, significantly enhancing the FPGA design process. For instance, executing a complete FPGA design flow for each design point is prohibitively expensive, and early-stage result estimations often lack the necessary accuracy to demonstrate the appropriate design trade-offs. A well-calibrated machine learning predictive model can replace such resource-intensive computations with a rapid approximation.
- **Flexible and versatile modeling.** Modern machine learning techniques, in contrast to traditional statistical data analysis methods, offer a broad spectrum of modeling options to accommodate the complex FPGA design processes. On one hand, machine learning provides various predictive formulations essential for addressing numerous FPGA design challenges, such as classification, clustering, regression, generative modeling, and more. On the other hand, contemporary machine learning approaches can manage versatile feature representations like graphs, circuit imaging, functional behaviors, etc., and learn intricate relationships between those features and target metrics.
- **Minimizing human supervision.** The application of machine learning in FPGA design reduces human supervision in the design process in two ways. Firstly, the conventional CAD tool R&D process heavily depends on expert knowledge in FPGA design and CAD algorithms, with most heuristics being developed through

extensive empirical efforts. Conversely, autonomous exploration and learning systems, such as reinforcement learning mechanisms, can substantially expedite the exploration process using an intelligent, self-guided agent.

A primary obstacle to rapid hardware specialization with FPGAs stems from weak guarantees of existing FPGA tools for achieving high-quality QoR [44–51]. To meet the diverse requirements of a broad range of application domains, current FPGA tools from academia [52] and industry [53, 54] provide a large set of options across multiple stages of the tool flow in the form of compile-/run-time directives, also known as `pragmas`. Due to the size and complexity of the design space spanned by these options, coupled with the time-consuming evaluation of each design point, deciding an optimal set of tool options, also known as *design space exploration* or DSE, has become remarkably challenging [55, 56]. To tackle this challenge, recently many ML-assisted frameworks have been proposed to automatically and intelligently decide on an optimal set of tool options to accelerate iterative QoR estimation. These frameworks utilize design-specific features extracted from the early stages of the design flow to guide the decision process with significant runtime savings [55, 56].

To meet stringent FPGA design objectives, e.g., area, latency, power, FPGA programmers need to explore a broad spectrum of FPGA customization options, known as Design Space Exploration (DSE). DSE is formulated as a multi-objective optimization problem. The result of the DSE is a set of Pareto-optimal FPGA designs. To customize the synthesis process to meet design objectives, almost all FPGA tools from academia [52] and industry [53, 54] provide numerous tunable knobs in the form of compiler directives, also known as `pragmas`. The usage of `pragmas` in the synthesis process results in a massive and convoluted design search space that is virtually impossible to explore manually or using an exhaustive search to identify one or more optimal combinations of `pragmas` corresponding to an optimal design point. In addition to that, how FPGA tools pre-characterize the area and delay of basic primitives (e.g., LUTs, BRAMs, DSPs), the results are inaccurate making it necessary to perform time-consuming logic synthesis after each newly generated design. Therefore there is an urgent need to devise automatic methods to quickly and efficiently explore design state space and identify an optimal combination of compiler directives.

There are multiple works that apply machine learning to automatically identify an optimal set of compile-time directives for C/C++/OpenCL-based designs at the HLS stage. Liu et al. [57] pose the DSE as a classification problem of identifying beneficial designs for synthesis. It incorporates pruning with an adaptive windowing method to find the candidate Pareto-optimal HLS designs. The adaptive windowing method is derived from the Rival Penalized Competitive Learning (RPCL) model using an important set of features (e.g., estimated area of a register, multiplexer, decoder, number of wires) adjusted on the fly during exploration. Transductive Experimental Design (TED) [58] aims to select a representative and hard-to-predict samples from the design space. The objective is to maximize the accuracy of the predictive model with the fewest possible training samples. TED assumes no a priori knowledge about the learning model and hence can be beneficial to any learning model.

Instead of improving the accuracy of the ML model, Adaptive Threshold Non-Pareto Elimination (ATNE) [59] primarily focuses on understanding and estimating the risk of losing good designs due to learning inaccuracy at the system level. Additionally, ATNE provides a Pareto identification threshold by adapting the estimated inaccuracy of the regressor for an efficient DSE. The work of [60] proposes a predictive model-based approach to finding meta-heuristics parameters (hyperparameters) of a multi-heuristic design space explorer consisting of simulated annealing, genetic algorithm, and ant colony optimization. To select an optimal combination of HLS directives, Lo et al. [61] incorporate low-fidelity estimates available from HLS tools in a multi-fidelity model and use a sequential model-based optimization [62] to explore the design space. To further enhance the sequential model, Lo et al. [63] use a hierarchical Gaussian process modeling to combine probabilistic estimates of component designs of a system to obtain exact values of system-level metrics, e.g., area, delay, and latency. Prospector [64] employs Bayesian techniques to optimize HLS synthesis `pragmas` to reduce execution latency and resource usage. Encoding the design space to capture design performance and FPGA costs (e.g., flip-flops, LUTs, BRAMs, DSPs) and sampling a small fraction (typically $<1\%$) of the design space to reveal optimal design are key to Prospector.

The authors in [65, 66] propose a transfer learning-based approach to transfer learned design space knowledge from source designs and apply it to a new target design. The key idea is multi-domain transfer learning in which effectively common knowledge between multiple source applications is extracted and is shared with the target applications. The objective is to enhance the training performance and reduce sample complexity.

Building upon transfer learning concepts, Wu et al. [67] recently presented IronMan, a combination of GNN [68] and reinforcement learning, which offers optimal solutions under user-defined constraints or a range of trade-offs (Pareto solutions) among various objectives, such as resources, area, and latency, for a given HLS C/C++ program. IronMan uncovers hidden optimization opportunities for increased parallelism and reduced latency, accurately predicts the performance of the generated RTL using only the original dataflow graph of the input program, consisting of both regular and irregular data paths, and examines optimal resource allocation strategies based on user-specified constraints.

In a separate line of research, ML is employed to determine the optimal set of compiler directives for HDL-based designs during logic synthesis or subsequent design stages. Kurek et al. [69] model an objective function using a Gaussian process and employ an SVM classifier to estimate if design constraints are met.

InTime [70] utilizes active learning for design space exploration (DSE) with ML models as a surrogate for actual design synthesis during design evaluation to identify a suitable combination of compiler directives. InTime constructs an ML model from a database of preliminary FPGA tool run results and predicts the next set of FPGA tool options to enhance timing results. To further refine the objective, InTime depends on a limited degree of statistical sampling. DATuner [71] is a parallel, iterative autotuning framework for FPGA compilation, using a multi-arm bandit technique to automatically select an appropriate set of compiler directives for a complete FPGA flow. DATuner employs dynamic solution space partitioning based on information (e.g., runtime, search quality) obtained from previous iterations and intelligently allocates computing resources to unexplored design subspaces and subspaces containing high-quality solutions. Mametjanov et al. [72] propose a model-based search framework that integrates sampling-based reduction of compiler directive space and guides the search toward promising directive configurations. LAMDA [73] takes an RTL description as input and automatically configures tool options across logic synthesis, placement, and routing stages. LAMDA circumvents iterating over the time-consuming FPGA implementation tool flow, particularly in place and route stages, by investigating potential speedups achievable by introducing high-fidelity QoR estimations in early and low-fidelity design stages, effectively pruning the large and complex search space early in the design flow. Contrary to previous approaches, LAMDA addresses the DSE problem from a multi-stage perspective, balancing the trade-off between computing effort and estimation accuracy.

9.6 Summary and Trends

Logic synthesis, as a critical step in high-level design information and physical implementation, directly influences the performance of FPGA chips. With the exponential growth in design scale and congestion issues in backend placement and routing, continuous research is required in logic representation, logic optimization, technology mapping, and design flow automation for FPGA logic synthesis.

The core challenge for logic synthesis tools is obtaining improved PPA within reasonable CPU time. In addition, different circuits demand distinct optimization strategies and tool flows, making effective exploration within the vast design space formed by numerous optimization commands within the tools a daunting task. Particularly, as process nodes evolve and design scale increases, addressing this challenge necessitates ongoing innovative solutions. Artificial Intelligence (AI) holds the promise of playing a more significant role in logic synthesis.

AI has demonstrated remarkable breakthroughs in various domains. Combining AI and EDA is an essential future trend, which includes:

1. intelligent scheduling within individual tools or optimization engines,
2. intelligent flows for EDA tools.

The integration of AI and EDA requires a substantial number of samples, thus necessitating further refinement of different logic representation and optimization methods in conventional synthesis for AI scheduling. One of the fundamental issues in logic synthesis is how to represent Boolean logic functions and develop optimization algorithms for relevant representations, as well as the usage of computational engines. For instance, methods like truth tables, SOP, BDD, and AIG are used to represent Boolean logic functions, and specific optimization algorithms are tailored for each method.

References

1. R.E. Bryant, Graph-based algorithms for Boolean function manipulation. Comput. IEEE Trans. **100**(8), 677–691 (1986)
2. A. Mishchenko, S. Chatterjee, R. Brayton, DAG-aware AIG rewriting: a fresh look at combinational logic synthesis, in *Design Automation Conference (DAC)* (2006), pp. 532–535
3. C. Yu, M.J. Ciesielski, M. Choudhury, A. Sullivan, Dag-aware logic synthesis of datapaths, in *Proceedings of the 53rd Annual Design Automation Conference, DAC 2016, Austin, TX, USA, June 5-9, 2016* (2016), pp. 135:1–135:6
4. C. Yu, M. Choudhury, A. Sullivan, M.J. Ciesielski, Advanced datapath synthesis using graph isomorphism, in *2017 IEEE/ACM International Conference on Computer-Aided Design, ICCAD 2017, Irvine, CA, USA, November 13-16, 2017* (2017) pp. 424–429
5. L. Amaru, P.-E. Gaillardon, G. De Micheli, Majority-inverter graph: a new paradigm for logic optimization. TCAD (2016)
6. L. Amarú, P.-E. Gaillardon, A. Chattopadhyay, G. De Micheli, A sound and complete axiomatization of majority-*n* logic. IEEE Trans. Comput. **65**(9), 2889–2895 (2015)
7. Z. Chu, W. Haaswijk, M. Soeken, Y. Xia, L. Wang, G. De Micheli, Exact synthesis of boolean functions in majority-of-five forms, in *2019 IEEE International Symposium on Circuits and Systems (ISCAS)* (IEEE, 2019), pp. 1–5
8. W. Haaswijk, M. Soeken, L. Amarú, P.-E. Gaillardon, G. De Micheli, A novel basis for logic rewriting, in *2017 22nd Asia and South Pacific Design Automation Conference (ASP-DAC)* (IEEE, 2017), pp. 151–156
9. Z. Chu, M. Soeken, Y. Xia, L. Wang, G. De Micheli, Structural rewriting in XOR-majority graphs (2019)
10. Z. Chu, Z. Li, Y. Xia, L. Wang, W. Liu, Bcd adder designs based on three-input XOR and majority gates. IEEE Trans. Circ. Syst. II: Express Briefs **68**(6), 1942–1946 (2020)
11. A.M.R. Brayton, Scalable logic synthesis using a simple circuit structure, in *Proceeding of IWLS*, vol. 6 (2006), pp. 15–22
12. W.V. Quine, The problem of simplifying truth functions. Am. Mathe. Monthly **59**(8), 521–531 (1952)
13. E.A. Ernst, *Optimal Combinational Multi-Level Logic Synthesis*. University of Michigan (2009)
14. N. Eén, Practical sat-a tutorial on applied satisfiability solving. Slides Invited Talk at FMCAD (2007)
15. D.E. Knuth, *The Art of Computer Programming*, vol. 3 (Pearson Education, 1997)
16. M. Soeken, L. G. Amaru, P.-E. Gaillardon, G. De Micheli, Exact synthesis of majority-inverter graphs and its applications. IEEE Trans. Comput.-Aid. Des. Integr. Circ. Syst. **36**(11), 1842–1855 (2017)
17. Z. Chu, M. Soeken, Y. Xia, L. Wang, G. De Micheli, Advanced functional decomposition using majority and its applications. IEEE Trans. Comput.-Aid. Des. Integr. Circ. Syst. **39**(8), 1621–1634 (2019)
18. C.E. Leiserson, J.B. Saxe, Retiming synchronous circuitry. Algorithmica **6**(1-6), 5–35 (1991)
19. A.P. Hurst, A. Mishchenko, R.K. Brayton, Fast minimum-register retiming via binary maximum-flow, in *Formal Methods in Computer Aided Design, 2007. FMCAD'07* (IEEE, 2007), pp. 181–187
20. A. Hurst, A. Mishchenko, R. Brayton, Scalable min-register retiming under timing and initializability constraints, in *Proceedings of the 45th Annual Design Automation Conference* (ACM, 2008), pp. 534–539
21. S.S. Sapatnekar, R.B. Deokar, Utilizing the retiming-skew equivalence in a practical algorithm for retiming large circuits. IEEE Trans. Comput.-Aided Des. Integrat. Circ. Syst. **15**(10), 1237–1248 (1996)
22. P. Pan, Continuous retiming: algorithms and applications, in *Computer Design: VLSI in Computers and Processors, 1997. ICCD'97. Proceedings., 1997 IEEE International Conference on.* (IEEE, 1997), pp. 116–121

23. A.V. Goldberg, An efficient implementation of a scaling minimum-cost flow algorithm. J. Algor. **22**(1), 1–29 (1997)
24. D.P. Singh, V. Manohararajah, S.D. Brown, Incremental retiming for FPGA physical synthesis, in *Proceedings of the 42nd annual Design Automation Conference* (ACM, 2005), pp. 433–438
25. S. Malik, E.M. Sentovich, R.K. Brayton, A. Sangiovanni-Vincentelli, Retiming and resynthesis: optimizing sequential networks with combinational techniques. IEEE Trans. Comput.-Aided Des. Integrat. Circ. Syst. **10**(1), 74–84 (1991)
26. G. De Micheli, Synchronous logic synthesis: algorithms for cycle-time minimization. IEEE Trans. Comput.-Aid. Des. Int. Circ. Syst. **10**(1), 63–73 (1991)
27. J. Cong, C. Wu, Optimal FPGA mapping and retiming with efficient initial state computation. IEEE Trans. Comput.-Aid. Des. Integr. Circ. Syst. **18**(11), 1595–1607 (1999)
28. N. Shenoy, Retiming: theory and practice. Int. VLSI J. **22**(1), 1–21 (1997)
29. A. Mishchenko, S. Chatterjee, R.K. Brayton, DAG-aware AIG rewriting a fresh look at combinational logic synthesis (2006), pp. 532–535
30. A. Mishchenko, R. Brayton, Recording synthesis history for sequential verification, in *Formal Methods in Computer-Aided Design, 2008. FMCAD'08.*(IEEE, 2008), pp. 1–8
31. C.E. Leiserson, F.M. Rose, J.B. Saxe, Optimizing synchronous circuitry by retiming (preliminary version), in *Third Caltech Conference on Very Large Scale Integration* (Springer, 1983), pp. 87–116
32. N. Shenoy, R. Rudell, Efficient implementation of retiming, in *Proceedings of the 1994 IEEE/ACM International Conference on Computer-Aided Design* (IEEE Computer Society Press, 1994), pp. 226–233
33. Z. Chu, M. Soeken, Y. Xia, G. De Micheli, Functional decomposition using majority, in *2018 23rd Asia and South Pacific Design Automation Conference (ASP-DAC)* (IEEE, 2018), pp. 676–681
34. H. Katebi, I. Markov, Large-scale boolean matching, in *Advanced Techniques in Logic Synthesis, Optimizations and Applications* (Springer, 2010), pp. 227–247
35. L. Machado, J. Cortadella, Support-reducing decomposition for FPGA mapping. IEEE Trans. Comput.-Aid. Des. Integr. Circ. Syst. **39**(1), 213–224 (2018)
36. A. Mishchenko, S. Chatterjee, R. Brayton, Improvements to technology mapping for LUT-based FPGAs, in *Proceedings of the 2006 ACM/SIGDA 14th International Symposium on Field Programmable Gate Arrays* (2006), pp. 41–49
37. D. Chen, J. Cong, DAOmap: a depth-optimal area optimization mapping algorithm for FPGA designs, in *IEEE/ACM International Conference on Computer Aided Design, 2004. ICCAD-2004* (IEEE, 2004), pp. 752–759
38. J. Cong, Y. Ding, On area/depth trade-off in LUT-based FPGA technology mapping, in *Proceedings of the 30th International Design Automation Conference* (1993), pp. 213–218
39. J. Cong, Y. Ding, On area/depth trade-off in LUT-based FPGA technology mapping. TVLSI (1994)
40. J. Cong, Y. Ding, FlowMap: an optimal technology mapping algorithm for delay optimization in lookup-table based FPGA designs. TCAD (1994)
41. J. Cong, Y.-Y. Hwang, Simultaneous depth and area minimization in LUT-based FPGA mapping, in *Proceedings of the 1995 ACM Third International Symposium on Field-Programmable Gate Arrays* (1995) pp. 68–74
42. K.R. Basireddy, S. Sabbavarapu, A. Acharyya, Cut-less technology mapping using shannon factor graph with on-the-fly size reduction. J. Low Power Electron. **14**(3), 448–457 (2018)
43. A. Mishchenko, S. Cho, S. Chatterjee, R. Brayton, Cutless FPGA mapping, ERL Technical Report, EECS Department, UC Berkeley, Technical Report (2007)
44. H. Ren, J. Hu, *Machine Learning Applications in Electronic Design Automation* (Springer Nature, 2023)
45. D. Pal, C. Deng, E. Ustun, C. Yu, Z. Zhang, Machine learning for agile FPGA design. Mach. Learn. Appl. Electron. Des. Automat. 471–504 (2022)
46. C. Yu, C.-C. Huang, G.-J. Nam, M. Choudhury, V. N. Kravets, A. Sullivan, M. Ciesielski, G. De Micheli, End-to-end industrial study of retiming. ISVLSI (2018)

47. C. Yu, Z. Zhang, Painting on placement: forecasting routing congestion using conditional generative adversarial nets. DAC (2019)
48. C. Yu, W. Zhou, Decision making in synthesis cross technologies using LSTMs and transfer learning. MLCAD (2020)
49. N. Wu, Y. Li, C. Hao, S. Dai, C. Yu, Y. Xie, Gamora: graph learning based symbolic reasoning for large-scale Boolean networks, in *ACM/IEEE Design Automation Conference (DAC'23)* (2023)
50. W.L. Neto, M.T. Moreira, L. Amaru, C. Yu, P.-E. Gaillardon, Read your circuit: leveraging word embedding to guide logic optimization. ASPDAC (2021)
51. W.L. Neto, M.T. Moreira, L. Amaru, C. Yu, SLAP: a supervised learning approach for priority cuts technology mapping. DAC (2021)
52. A. Canis, J. Choi, M. Aldham, V. Zhang, A. Kammoona, J.H. Anderson, S. Brown, T. Czajkowski, LegUp: high-level synthesis for FPGA-based processor/accelerator systems. FPGA (2011)
53. J. Cong, B. Liu, S. Neuendorffer, J. Noguera, K. Vissers, Z. Zhang, High-level synthesis for FPGAs: from prototyping to deployment. TCAD (2011)
54. Intel HLS Compiler. https://www.intel.com/content/www/us/en/software/programmable/quartus-prime/hls-compiler.html. Accessed: 08 Nov 2023 14:14:07.
55. C. Yu, Flowtune: practical multi-armed bandits in Boolean optimization, in *International Conference On Computer Aided Design (ICCAD)* (IEEE, 2020), pp. 1–9
56. W.L. Neto, Y. Li, P.-E. Gaillardon, C. Yu, Flowtune: end-to-end automatic logic optimization exploration via domain-specific multi-armed bandit. IEEE Trans. Comput.-Aid. Des. Integr. Circ. Syst. (2022)
57. D. Liu, B.C. Schafer, Efficient and reliable high-level synthesis design space explorer for FPGAs. FPL (2016)
58. H.-Y. Liu, L.P. Carloni, On learning-based methods for design-space exploration with high-level synthesis. DAC (2013)
59. P. Meng, A. Althoff, Q. Gautier, R. Kastner, Adaptive threshold non-pareto elimination: re-thinking machine learning for system-level design space exploration on FPGAs. DATE (2016)
60. Z. Wang, B.C. Schafer, Machine learning to set meta-heuristic specific parameters for high-level synthesis design space exploration. DAC (2020)
61. C. Lo, P. Chow, Multi-fidelity optimization for high-level synthesis directives. FPL (2018)
62. C. Lo, P. Chow, Model-based optimization of high-level synthesis directives. FPL (2016)
63. C. Lo, P. Chow, Hierarchical modelling of generators in design-space exploration. FCCM (2020)
64. A. Mehrabi, A. Manocha, B.C. Lee, D.J. Sorin, Prospector: synthesizing efficient accelerators via statistical learning. DATE (2020)
65. J. Kwon, L.P. Carloni, Transfer learning for design-space exploration with high-level synthesis. MLCAD (2020)
66. M. Kurek, M.P. Deisenroth, W. Luk, T. Todman, Knowledge transfer in automatic optimisation of reconfigurable designs. FCCM (2016)
67. N. Wu, Y. Xie, C. Hao, IronMan: GNN-assisted design space exploration in high-level synthesis via reinforcement learning. GLSVLSI (2021)
68. W. Hamilton, Z. Ying, J. Leskovec, Inductive representation learning on large graphs. NIPS (2017)
69. M. Kurek, T. Becker, T. C. Chau, W. Luk, Automating optimization of reconfigurable designs. FCCM (2014)
70. N. Kapre, H. Ng, K. Teo, J. Naude, InTime: a machine learning approach for efficient selection of FPGA CAD tool parameters. FPGA (2015)
71. C. Xu, G. Liu, R. Zhao, S. Yang, G. Luo, Z. Zhang, A parallel bandit-based approach for autotuning FPGA compilation. FPGA (2017)

72. A. Mametjanov, P. Balaprakash, C. Choudary, P.D. Hovland, S.M. Wild, G. Sabin, Autotuning FPGA design parameters for performance and power. FCCM (2015)
73. E. Ustun, S. Xiang, J. Gui, C. Yu, Z. Zhang, LAMDA: learning-assisted multi-stage autotuning for FPGA design closure. FCCM (2019)

Chapter 10
Physical Implementation

Abstract In this chapter, physical implementation step of FPGA application design will be investigated, including the well-known EDA steps: packing, placement, and routing. Packing cluster the FPGA atoms together into larger design units; placement assign each design unit to a proper location on the device; routing finds the optimized wire path to connect all these design units.

10.1 Packing

Packing is the first and one of the critical steps when implementing a post-synthesis netlist on a given FPGA, as depicted in Fig. 10.1. Packing algorithm aims to cluster logic primitives, e.g., Look-Up Tables, flip-flops, etc., and efficiently map them to logic blocks in an FPGA. Staying at the most upstream of the implementation flow, the result of packing will profoundly impact the placement, and routing results. Its capability of foreseeing critical paths, placement restriction, and routing congestion can significant improve the overall performance of placement and routing. Therefore, packing algorithms have been intensively researched since the born of FPGA devices.

In this part, we will first define the problem of packing algorithms to solve, and then introduce two well-known types of packing algorithms, and also discuss future trends in this field.

10.1.1 Overview

In sophisticated implementation framework, e.g., the one shown in Fig. 10.2, the input and output data structures of packing algorithms are typically well defined and even standardized (See details in Chap. 3). Figure 10.2 depicts a detailed flow chart for all the existing packing algorithms. A packing algorithm converts a post-synthesis netlist to a clustered netlist, with some optional inputs such as a detailed FPGA device modeling. The clustered netlist is the input of downstream engines, i.e., a placement algorithm.

K. Tu et al., *FPGA EDA*, https://doi.org/10.1007/978-981-99-7755-0_10

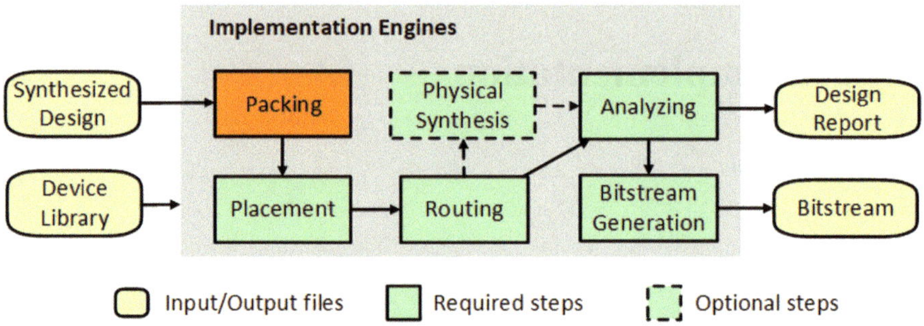

Fig. 10.1 Packing's position in FPGA application EDA flow

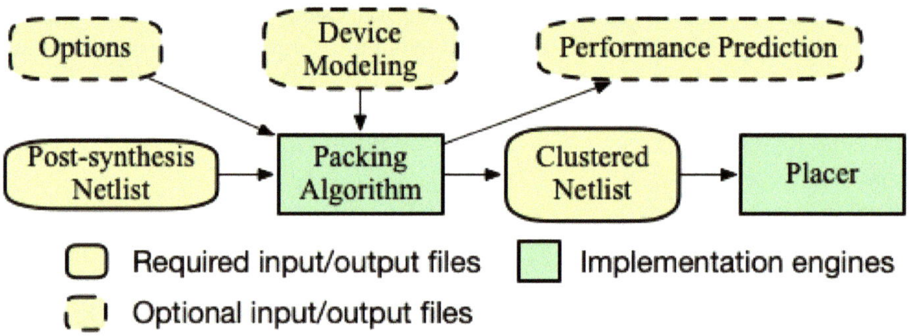

Fig. 10.2 A typical EDA flow for a packing algorithm: input and output data structures

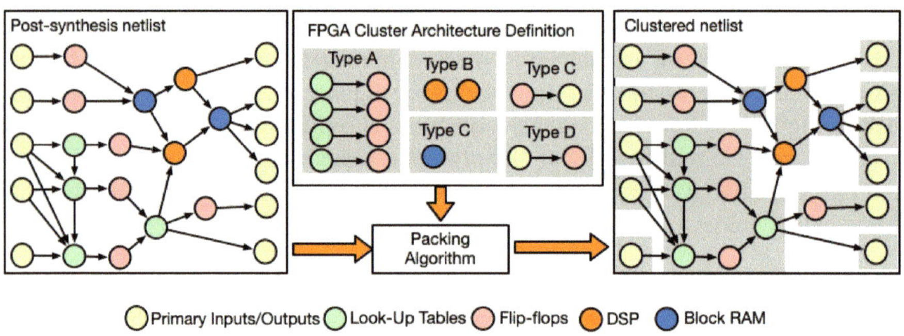

Fig. 10.3 An illustration on the problem definition of FPGA packing algorithms. Each gray box in the clustered netlist denotes a cluster consisting of several primitives

Problem Definition: As an intermediate step between logic synthesis and placements, packing algorithms are designed to cluster and map atom-level logic primitives to dedicated logic resources in programmable blocks, e.g., *Configurable Logic Block* (CLB). (Fig. 10.3) explains the problem definition of packing algorithms through an illustrative example. Packing algorithms require two inputs:

Table 10.1 Comparison on seed-based and partition-based packing algorithms

Metric	Seed-based			Partition-based	Hybrid
Representative tools	TVPack [1, 2]	RPack [3, 4]	AAPack [5, 6]	PPack [7, 8]	HDPack [9]
Support flexible CLB arch.	×	×	✓	×	×
Quality-of-result optimality	Local	Local	Local	Global	Global
Time complexity	Low	Low	Low	High	Medium
Support heterogeneous blocks	×	×	✓	×	×
Open-source	✓	×	✓	×	×

1. A post-synthesis netlist, generated by synthesis tools (See Chap. 9 for details), which consists of a logic network of primitives, such as *Look-Up Tables* (LUTs), *Flip-Flops* (FFs), *Digital Signal Processing* (DSP), *Random Access Memory* (RAM) *etc.* These logic primitives are defined in the technology library of logic synthesis tools.
2. An FPGA cluster architecture description, which describes detailed architectures of programmable blocks in FPGA tiles, such as CLB, DSP, BRAM, and I/O. The technical details include the number of primitives per block, the number input and outputs per block, and also the programmable routing architecture down to pin-to-pin connections. To model the architecture details, packing tools typically build a graph, where nodes represent primitives and edges denote routing resources.

Based on the two inputs, packing algorithms output a clustered netlist which only consists of a number of clusters. Each cluster includes

1. a legal placement of primitives onto logic resources;
2. a legal assignment of nets of the placed primitives to inputs and outputs of clusters;
3. a legal routing of the nets which drives or are driven by the primitives in the programmable block.

Each cluster will be treated as an individual block to be placed for placement algorithms (see details in Sect. 10.2). The nets mapped to inputs and outputs of a cluster will be treated as source and sink nodes for routing algorithms (see details in Sect. 10.3).

In general, to evaluate the quality of packing algorithms, the *Power, Performance, Area* (PPA) at post-routing stage (see Fig. 10.1) are the golden metrics. However, some other metrics are also used to early predict PPA after packing is accomplished:

1. Number of clusters, which indicates the capability of algorithms to group primitives. To maximize the resource utilization of FPGA devices, packing algorithms should result in the fewest number of clusters. Each FPGA device contains a fixed number of programmable blocks for each type (see example in Fig. 10.3. If the

number of clusters exceeds their limits, placement and routing will definitely fail. As a result, a HDL design can not be implemented on the given FPGA device.
2. Number of nets, which indicates the number of interconnections between clusters. The minimize the routing congestion during placement and routing, packing algorithms should absorb as many net as possible into clusters. Any nets remains outside clusters will have to be routed by routing algorithms in downstream stage. FPGA devices contains a limited number of routing resources. A large number of nets may potentially cause overuse of routing resources, leading to failures in routing stage.

Mainstream Algorithms: Depending on the FPGA architecture, packing strategies can be very different. Mainstream packing algorithms can be categorized into three classes: seed-based, partition-based, and a hybrid of the previous two. Table 10.1 compares critical features between the different types of algorithms and their representative tools.

1. The seed-based algorithms are most widely researched and published in the past decade [1–6, 10, 11]. The seed-based algorithms highly rely on a cost function to guide the clustering engine, in order to optimize P.P.A.. The cost function has been intensively studied, which result in various algorithms/tools. Due to the bottom-up nature, the algorithms often hit a local optimal in clustering results. However, thanks to their simplicity and flexibility, the seed-based algorithms are default packing algorithm *AAPack* in the well-known academia FPGA architecture exploration tool *Verilog-to-Routing* [6, 12]. AAPack can now supports highly flexible CLB architectures as well as DSP, BRAM blocks, which are ubiquitous in modern FPGAs devices.
2. The partition-based packing algorithms utilize the graph partition algorithms, e.g, hMetis [13], to produce initial packing results and apply refinement to legalize each cluster. When compared to seed-based algorithms, the partition-based algorithms are more time-consuming (10×) due to the use of graph partitioner [7, 8]. However, partition-based algorithms follow a top-down optimization strategy, which can achieve global optimal results. On average, it can improve P.P.A. by 12% and routability by at 32% when compared to seed-based algorithms.
3. To combine the benefits of seed-based and partition-based algorithms, hybrid algorithms are proposed to perform partitioning as a coarse placement and then annotate the predicted physical location to the cost function of seed-based clustering. On average, it can improve P.P.A. by 7% and routability by at 25% when compared to seed-based algorithms. As the algorithm still rely on seed-based clustering, its timing complexity is similar to the seed-based algorithms with a limited overhead (within 13%).

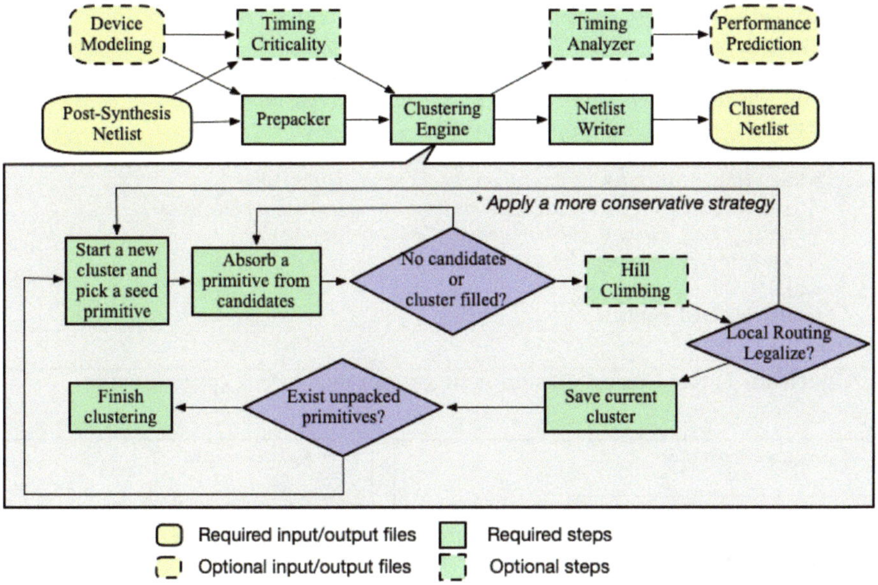

Fig. 10.4 A flow chart to illustrate seed-based packing algorithms

10.1.2 Seed-Based Packing Algorithms

As a mainstream type of packing algorithms, the seed-based packing algorithms have been well studied in the past decades. Even though there are more than ten variants of the seed-based algorithms, the principles are all the same, as illustrated in Fig. 10.2. In this section, we explain the algorithms based on a widely used implementation, i.e., *Architecture-Aware Pack* (AAPack) [6]. Figure 10.4 illustrates the algorithm of AAPack, which is a super set of other seed-based algorithms. The clustering engine of AAPack follows a greedy approach as other seed-based algorithms [1–4]. Clustering is applied to logic blocks, e.g., *Configurable Logic Block* (CLB) *etc.*, in a one-by-one manner, as depicted in Algorithm 1. A new cluster is created with a selected seed, expanded by absorbing primitives through a cost function, and ended when the cluster is fully filled or no more primitives can be absorbed. Each cluster is signed off by a legalizer to guarantee its resource utilization is below a threshold, e.g., 100%. Once a cluster is marked as packed, it will not be revisited by other clusters and no modifications are allowed. As AAPack is designed to be adaptive to various FPGA architectures and offer optimization in different objectives, it contains several unique features (as highlighted by green boxes using dash lines in Fig. 10.4: pre-packing, timing criticality computation, and hill climbing. In the rest of the part, we will present the algorithm details about each technical features.

Pre-pack: To maximize the routability, logic blocks in modern FPGAs typically contains highly flexible routing architectures. However, to efficiently implement

atom_netlist: *Post synthesis netlist*
architecture: *Device modeling showing detailed logic block and routing architectures.*
clustered_netlist: *Outputted netlist which contains clusters only.*
current_cluster: *Current cluster which is open for clustering primitives.*

```
Function pack(atom_netlist, architecture):
    clustered_netlist = empty;
    while exist_unpacked_candidates(atom_netlist) do
        current_cluster = open_new_cluster(atom_netlist, architecture);
        try_fill(current_cluster, atom_netlist, architecture);
        clustered_netlist.append(current_cluster);
    end
    return clustered_netlist;
end
```

Algorithm 1: Seed-based clustering algorithm in AAPack (pseudo-code)

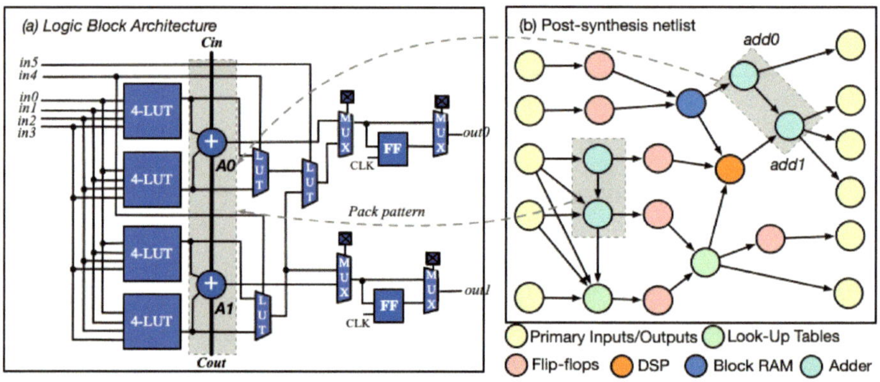

Fig. 10.5 A illustrative example for pre-packing: force packing patterns on post-synthesis netlists based on hard adder chains in a logic block

some frequently used logic functions, such as adder, modern FPGAs also includes a small portion of inflexible routing architectures. These architectures may cause complications in seed-based packers because the packers lack necessary information to map primitives so that the inflexible routing architectures can be satisfied. The most common inflexible routing architecture is on adder chains, as depicted in Fig. 10.5a. The mapping of primitives on the chain has to be exact without any flexibility in location, due to the hard wires in the chain. For example, the two adders `add0` and `add1` in a post-synthesis netlist of Fig. 10.5b, have to be mapped to the logic resource `A0` and `A1` in a cluster, respectively. Therefore, a pre-packing stage is required to extract such patterns of primitives from post-synthesis netlists and force a high priority when mapping to a cluster. As a result, the primitives in a pattern should be added to a cluster only as a group, to avoid potential failures seen in legalizers due to the strong limitation in routing architecture. Algorithm 2 presents the pseudo codes of the pre-packing stage. In the first step, a list of pack patterns

```
<!-- Pack pattern is forced on hard interconnect -->
<direct type="direct2" input="A1.cin" output="A0.cout"/>
 <pack_pattern name="carry_chain" in_port="A0.cout"
out_port="A1.cin"/>
</direct>
```

Fig. 10.6 Examples of pack pattern definition in FPGA architecture language

is extracted from the logic block architectures, where the types of primitives that are mappable to the patterns are identified. Note that pre-packing requires users to define pack patterns when crafting their FPGA architectures. Figure 10.6 shows an example of pack pattern for the hard adder chain in Fig. 10.5a, using the *University of Toronto FPGA Architecture Language* (UTFAL) [14]. In the second step, primitives in a post-synthesis netlist are grouped to molecules, which are the smallest unit to be considered/added to a cluster. Note that a molecule may consist of multiple primitives, such as a adder chain, or a single primitive, e.g., a LUT. When traversing the netlist to form molecules, the pre-packer starts with the largest pack pattern and ends with the smallest pack pattern. During each traversal, the pre-packer considers a pack pattern, and try to match parts of the netlist to a pack pattern through a technique similar to the matching process in a standard cell technology mappers [15]. The molecule creation process is greedy to ensure that each primitive can only be assigned to molecule.

atom_netlist: *Post synthesis netlist*
architecture: *Device modeling showing detailed logic block and routing architectures.*
molecule_list: *Group of primitives which are mappable to logic resources.*
current_cluster: *Current cluster which is open for clustering primitives.*
Function `pre-pack` (*atom_netlist, architecture*):
  pack_patterns = `extract_pack_patterns` (*architecture*);
  molecule_list = `build_molecule_list` (*atom_netlist, pack_patterns*);
  return pack_patterns, molecule_list;
end

Algorithm 2: Pre-packing algorithm in AAPack (pseudo-code)

Timing Criticality Computation: Timing-driven packing algorithms require timing criticality as a key factor in their cost functions, when select a molecule from candidates to add to a cluster. Therefore, a static timing analysis is run on the post-synthesis netlist, and timing criticality is computed for each net of molecules before clustering stage. Timing parameters of each primitive are annotated from the device modeling, including pin-to-pin delays, setup, and hold time. The timing analyzer predicts the routing delay between molecules based on the segment delays defined in routing architecture of device modeling. For example, AAPack assumes all length-4 wires used to interconnect molecules, which is pessimistic. Through a timing analysis, the slack of each edge is computed, based on which the timing criticality of each edge is achieved. Figure 10.7a shows an example of how a post-synthesis netlist is

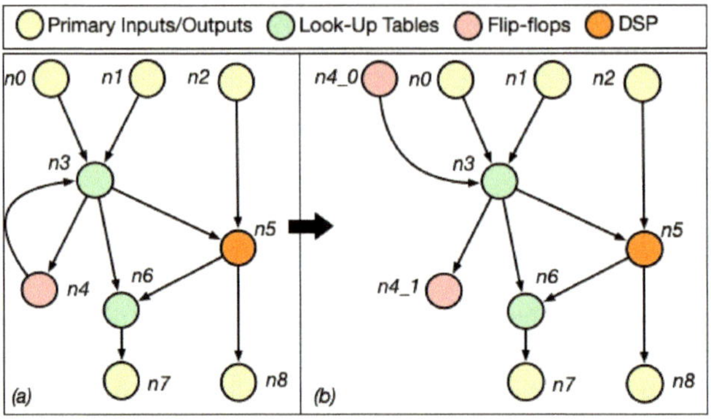

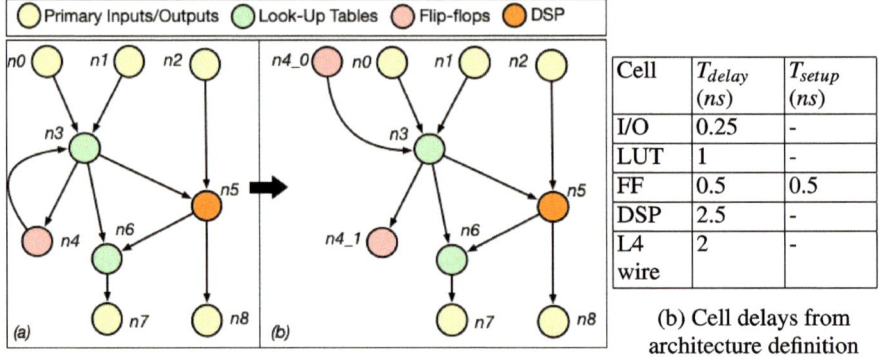

(a) An example of timing graph in the context of packer: (a) netlist and (b) timing graph

Cell	T_{delay} (ns)	T_{setup} (ns)
I/O	0.25	-
LUT	1	-
FF	0.5	0.5
DSP	2.5	-
L4 wire	2	-

(b) Cell delays from architecture definition

Fig. 10.7 An example of computing timing criticality in packing

Table 10.2 Timing criticality for each node in Fig. 10.7, where nodes on critical path is highlighted

Node	T_{arrival} (ns)	T_{required} (ns)	Slack	$T_{\text{criticality}}$
$n0$	0.25	0.5	0.25	0.98
$n1$	0.25	0.5	0.25	0.98
$n2$	0.25	6	5.75	0.58
$n3$	**3.5**	**3.5**	**0**	**1**
$n4\_0$	**0.5**	**0.5**	**0**	**1**
$n4\_1$	6	13.25	7.25	0.47
$n5$	**8**	**8**	**0**	**1**
$n6$	**11**	**11**	**0**	**1**
$n7$	**13.25**	**13.25**	**0**	**1**
$n8$	10.25	13.25	3	0.78

converted to a timing graph. Note that a timing path can start from either a primary input or an output of a flip-flop, and end to either a primary output or an input of a flip-flop. Therefore, the node $n4$ is splitted into two nodes $n4\_0$ (as the starting point of a timing path) and $n4\_1$ (as the ending point of a timing path) in the resulting timing graph of Fig. 10.7a. Based on the timing parameters listed in Fig. 10.7b, there is one critical path in the timing graph: $n4\_0 \rightarrow n3 \rightarrow n5 \rightarrow n6 \rightarrow n7$. The critical path delay is $13.75ns$. The timing criticality of each node can be computed by following Eq. 10.1, as detailed in Table 10.2. Note that all the nodes on the critical path have timing criticality of one. By assigning high timing criticality to the nodes on critical paths and close to critical paths, packers can be guided to optimize their timing using cost functions (see details in next part). Typically, seed-based packers cluster these timing critical nodes into the same logic block, in order to minimize the final critical path delay.

$$
\begin{aligned}
\text{slack}_{\text{node\_}i} &= T_{\text{required, node\_}i} - T_{\text{arrival, node\_}i} \\
\text{timing\_criticality}_{\text{node\_}i} &= 1 - \frac{\text{slack}_{\text{node\_}i}}{\text{crit\_path\_delay}}
\end{aligned}
\tag{10.1}
$$

```
molecule_list: Molecules extracted from post synthesis netlist
architecture: Device modeling showing detailed logic block and routing architectures.
current_cluster: Current cluster which is open for clustering primitives.
Function TryFill(current_cluster, molecule_list, architecture):
    candidates = update_candidate_list_for_cluster(current_cluster,
      molecule_list);
    while is_empty(candidates) or !is_cluster_full(current_cluster) do
        chosen_candidate = pick_molecule_with_highest_gain(current_cluster,
          candidates);
        try_place_molecule_in_cluster(current_cluster, chosen_candidate);
        if cluster_is_legal(current_cluster, architecture) then
            save_cluster_result(current_cluster);
            update_candidate_list_for_cluster(current_cluster,
              molecule_list);
        end
        else
            remove_molecule_from_candidates(chosen_candidate, candidates);
        end
    end
    if !is_cluster_full(current_cluster) then
        try_hill_climb(current_cluster);
    end
end
```

Algorithm 3: Detailed algorithm of the TryFill() function show in Algorithm 1 (pseudo-code)

Clustering engine: Seed-based packers build clusters one at a time, as explained in Algorithm 1. During clustering, molecules, created in the pre-packing stage, are grouped to form a cluster, and mapped to specific locations in the clusters. There are two phases when building a cluster:

1. *Seed selection*: Seed molecule is the first element that is added to a new cluster. Packers pick a molecule based on a cost function, which is considered to bring highest gain to the cluster. The type of gain depends on the objective in optimization, which can be routability-driven or timing-driven, *etc.*. To reach the designated goal, cost functions are built to quantify the gain of each molecule when added to a cluster. Therefore, unclustered molecules can be ranked by the gain, and packers pick a high-ranking molecule as the seed. Table 10.3 lists the seed selection strategy of representative packers. Take the example of VPACK, the molecule `m0` in Fig. 10.8a is selected as the seed when creating a cluster, because it has the most used inputs (= 4).

2. *Cluster filling*: Once a seed is picked, clustering engine fills a new cluster through iterations. Algorithm 3 summarizes the principles of the iterative cluster-filling approach from all the existing seed-based packers. In each iteration, a list of candidate molecules is created based a cost function which quantifies their gain to the cluster. The candidates may be ranked by the highest gain, and the packer can fastly spot the preferred molecule when added it to the cluster. This methodology is similar to the seed selection phase. With a list of the candidates, clustering engine pick a molecule in each iteration and try to place it onto an available location in the cluster. Considering the example in Fig. 10.8a, the molecule `m3` will be placed to the cluster, due to its highest gain than the others `m2` and `m4`. Note that absorbing a candidate with highest gain may not always lead to a success addition. A legalizer is called to ensure that (a) the new member will not exceed the input and output bandwidth of the logic block architecture, (b) when there are local architecture, all the input and output signals can be routed inside the cluster. Take the example in Fig. 10.8a, the molecule `m3` should be legally added to the cluster, as shown in Fig. 10.8b, when the input bandwidth is 6. However, if the input bandwidth is 5, the legalizer will fail when adding the molecule `m3` to the cluster. Instead, the molecule `m2` will be added to the cluster legally. If the legalizer passes, the clustering results, including the placement of the new member and routing traces, are saved and the candidate list will be updated for the next iteration. As a member have been added to the cluster, the gain of unclustered molecules are no longer the same as previous iteration. If the legalizer fails, the candidate will be removed from the list, and will not be considered in future iterations. This phase ends until there are no more candidates or the current cluster is full. To ensure highest gain in each iteration, the selected candidates should have direct connections to the cluster, being either a fan-in or a fanout molecule. As a result, a cluster may not be fully filled due to a limited number of candidates. A hill climbing phase may be called to fill the cluster. Hill climbing aim to increase the resource utilization rate, which considers candidate molecules which have not connections to the cluster. Consider the example in Fig. 10.8c, if there is an input bandwidth of 5, molecule `m3` cannot be added the cluster since it will result in 6 inputs for the cluster. In such case, hill climbing can add an unrelated molecule `m4` to the cluster without violating input bandwidth. Details about hill climbing can be found in [1].

Cost functions: The key factor in seed-based clustering engines is the cost functions, which profoundly impacts the decision in each iteration. Cost functions are considered in both seed selection and cluster-filling phases. Cost functions are design to reflect the affinity between molecules, indicating the gains to absorb a molecule into an existing cluster. A well designed cost function can precisely quantify the gain for each candidate molecule. Provide a rank which clearly distinguish the molecule with highest gain from the others. Therefore, with a sophisticated cost function, a clustering engine can make correct decision in each iteration toward the optimization goal. In contrast, a poor cost function may probably mislead the clustering engines to absorb a improper candidate, being far from the objective, e.g., maximum operating frequency. Depending on the applications, different optimization targets are

Table 10.3 Cost functions in representative packers

Packer	Type	Seed selection phase	Cluster filling phase
VPack [1]	Routability	$\max\{used\_inputs\}$	$\lvert\text{Nets}(M) \bigcap \text{Nets}(C)\rvert$
TVPack [2]	Routability/timing	Same as VPack	$\alpha \cdot TC(M) + (1-\alpha) \cdot \frac{\lvert Nets(M) \bigcap \text{Nets}(C)\rvert}{G}$
RPack [3]	Routability	Same as VPack	$\sum_{i \in \text{Nets}(M)} g(i, \text{Nets}(C), M)$
iRAC [4]	Routability	separation/degree^2	$2 \cdot n \cdot w(M) \cdot (1 + \alpha_{M,C})$
AAPack [5]	Routability/timing	$\alpha \cdot TC + (1-\alpha) \cdot$ used_inputs	$\alpha \cdot \text{TC} + (1-\alpha) \cdot$ connection_gain

M denotes a molecule candidate, while C denotes a cluster
TC denotes the timing criticality in Eq. 10.1

given to clustering engines, resulting in a different choice on cost functions. The two most studied goals are the routability-driven and the timing-driven, respectively. Table 10.3 compares the cost functions used by representative packers in the seed selection phase and cluster filling phase. Early works, such as VPack, RPack and iRac, employ single-objective cost functions and focus on routability optimization [1, 3, 4]. Therefore, the cost functions aim to estimate the number of nets to be absorbed into a cluster. VPack uses the number of common nets between a cluster and a candidate molecule. However, the cost function in VPack is too simplified to optimize routability, especially when there are high-fanout nets. It may mislead packer to prioritize absorbing molecules with common input nets, missing the opportunity to fully absorb low fanout nets. RPack, and iRAC's cost functions focus on fully absorbing nets, and minimize the number of external nets outside a cluster, by leveraging the Rent's rule. For example, in Fig. 10.8a, VPACK will assign the same gain to the candidate molecules `m2` and `m3`, and it may decide to cluster `m3` as depicted in Fig. 10.8b. However, the cost function of RPack and iRAC will assign a higher gain to `m2` than `m3`, because it can reduce more external nets. When comparing Fig. 10.8b, c, the cluster with `m2` contains 5 input nets and 1 output nets, while the cluster with `m3` contains 6 input nets and 1 output nets. Timing-driven packers, such as TVPack and AAPack, introduce the timing criticality into the cost function, as well as a factor α to balance the weights between timing gain and connection gain. The factor α is typically an empirical number achieved by experiments on a set of benchmarks, w.r.t. the best overall performance. When $\alpha = 1$, the cost function is biased to timing critically, resulting in a fully timing-driven clustering strategy. When $\alpha = 0$, the cost function is biased to routability, resulting in a fully connection-driven clustering strategy. We refer interested readers to [1–5] for further details.

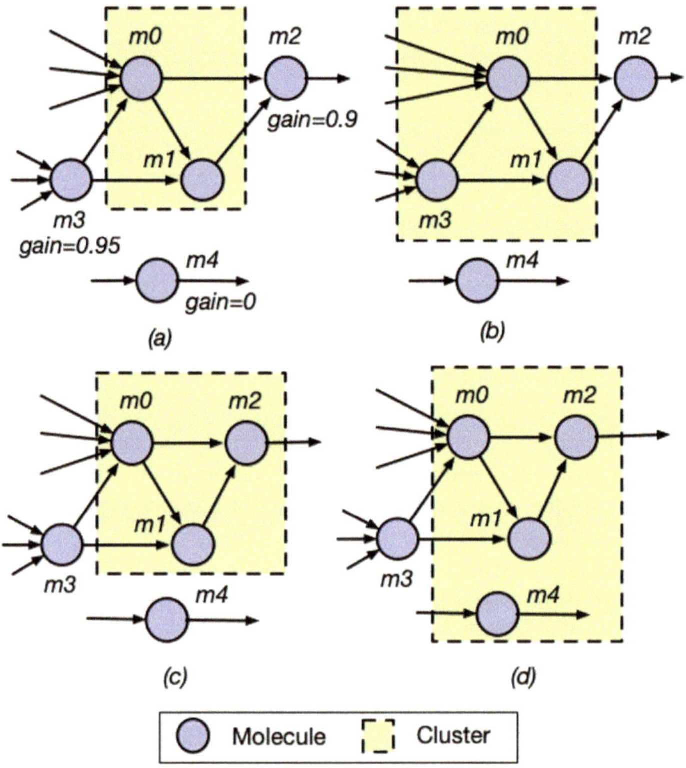

Fig. 10.8 A illustrative example for clustering: **a** an example cluster; **b** absorb the molecule with highest gain; **c** absorb the molecule subject to input bandwidth; **d** hill climbing example

10.1.3 Partition-Based Packing Algorithms

Though seed-based packing algorithms are fast in runtime to produce tight clusters and easy to be tuned for various constraints, they may become stuck in local minima due to the localized greedy strategy. Since clusters are built one by one and clusters cannot be modified once built, it is difficult for seed-based approaches to further optimize QoR from a global view. To compensate the loss, cluster modification, such as *Basic Logic Element* (BLE) swapping can be applied in downstream placement stages [16]. However, such fix is expensive in runtime, which may cause massive changes to clustering results, defeating the initial purpose of packing. Therefore, partition-based packing algorithms are proposed to produce high-quality results through a global optimization approach [7, 8].

To avoid local optimal, partition-based algorithms aim to form clusters with a global view, starting from the first step. Algorithm 4 presents the major steps for partition-based packers [7, 8, 17]. There are three key factors which impacts most on the QoR of the partition-based packers:

```
atom_netlist: Post synthesis netlist
architecture: Device modeling showing detailed logic block and routing architectures.
clustered_netlist: Outputted netlist which contains clusters only.
Function pack(atom_netlist, architecture):
    hypergraph = convert_netlist_to_a_hypergraph(atom_netlist);
    if timing_driven then
        add_weighted_timing_edges_to_hypergraph(hypergraph);
    end
    partitions = kway_partitioner(hypergraph);
    fine_tune_partitions_under_constraints(partitions, architecture);
    clustered_netlist = convert_partition_results_to_clusters(partitions);
    return clustered_netlist;
end
```

Algorithm 4: Partition-based clustering algorithm in PPack/TPPack [7] (pseudo-code)

1. an efficient graph partitioner, which can partition a large graph evely into small groups within a limited amount of runtime. For example, hMetis can produce extremely high-quality bisections of hypergraphs with 100,000 vertices in well under 3 min on an R10000-based SGI workstation and a Pentium Pro-based personal computer [18].
2. a proper modeling of the input netlists `atom_netlist` in hypergraph, which can be accepted by graph partitioners. In particular, the choice of nodes and the weight of edges have profound and direct impacts on the decision of partitioners. For different FPGA architectures, the node and weight build-up may be very different, in order to guide graph partitioner with a clear focus on optimization.
3. rebalancing partitions subject to design constraints. While current state-of-art graph partitioners focus on optimizing connectivity, generated partitions may contain a large of nodes, exceeding the logic capacity of a BLE in FPGA. The rebalancing aims to fixes up these exceptions by moving nodes from oversized partitions to undersized.

In the rest of this part, technical details about each critical step of the partition-based algorithms are presented.

Hypergraph conversion: Hypergraph is a standard and general-purpose graph network which are frequently used by graph partitioners as inputs [19]. Post-synthesis netlists are typically modeled as logic networks, as illustrated in Fig. 10.9a, where each node represents a LUT or a FF or an I/O. When converting such logic network to a hypergraph, a pre-packing step is required, being similar to the seed-based packers (see details in 10.1.2). During pre-packing, LUTs and FFs are paired to become super nodes. Pre-packing not only efficiently reduces the graph size and hence the runtime of graph partitioning, but also avoids the exposure of unnecessary edges which may mislead partitioners. For example, the nodes representing LUTs and FFs `{n0, n1, ..., n7}` in Fig. 10.9b are grouped into BLE nodes `{b0, b1, b2, b3}` in Fig. 10.9c. Note that there are tight connections inside each BLE between

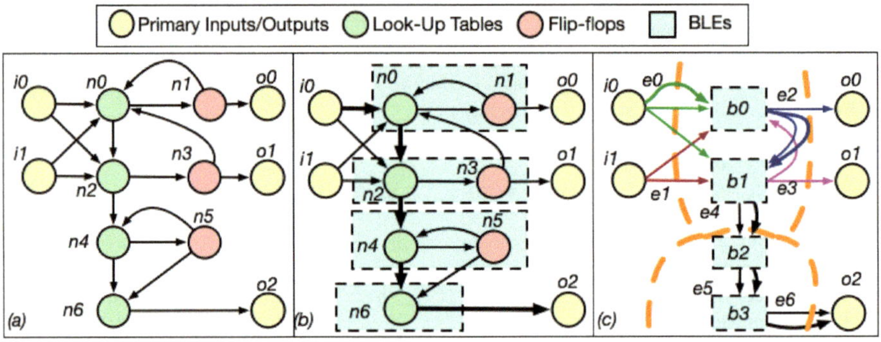

Fig. 10.9 A illustrative example for netlist to hypergraph conversion: **a** an example of input netlist; **b** group nodes into BLEs; **c** hypergraph representation with a potential partitioning

LUTs and FFs. Even these connections are shown in the hypergraph, they may change the partition results with a small probability, as partition algorithms are designed to minimize connectivity. However, as the time complexity of graph partition algorithms is typically high, hiding these details can reduce runtime while have almost no impacts on results. In addition, edges in the logic network are combined to hyperedges between nodes. For example, the edges sharing the same source node `i0` in Fig. 10.9b become a hyperedge `e0` in Fig. 10.9c. We refer interested readers to [19] for concepts of hypergraph.

Timing-driven edge addition: Timing-driven is a critical feature required by modern FPGA design tools. Partitioning-based packers can be made timing-driven by adding a number of extra edges to a hypergraph. This encourages graph partitioner to group the two-end nodes of these edges into the same partition, since the costs of these edges are significantly higher than others. The weight adjustment is done by two step:

1. Identify timing critical edges. The timing criticality or slack of each edge in an input netlist can be computed by the static timing analysis, similar to seed-based packers.
2. For each timing critical edge, add a weighted edge between the two corresponding nodes to the hypergraph. For example, the edges on a critical path (marked by bold lines) in Fig. 10.9b are added to the resulting hypergraph, as show in `{e0, e2, e4, e5, e6}` in Fig. 10.9c.

Note that different packers may have different methods when converting timing criticality to edge weights. For example, for each edge e_i, PPack considers both the pack-accumulated timing weight $pw(e_i)$ and the timing slack slack(e_i):

$$w(e_i) = \alpha \cdot \left(1 - \frac{\text{slack}(e_i)}{\text{slack}_{\max}}\right) + (1 - \alpha) \cdot \frac{pw(e_i)^{\beta}}{pw_{\max}} \tag{10.2}$$

where $\text{slack}_{\max}$ and $pw_{\max}$ denote the largest slack and pack-accumulated timing weight among all the edges, respectively. α is the weight factor for slack-based criticality, similar to the weight factor used in TVPack and AAPack shown in Table 10.3. β is an exponent (smaller than 1) to bring all $pw(e_i)$ values to a more comparable level related to $pw_{\max}$ (otherwise, most of them would be near 0). When adding timing edges to a hypergraph, the weight of each edge is scaled to an integer through $[M \cdot w(e_i)]$, where M is the upper bound. In practice, the number of timing edges to be added may be limited by an upper bound p (a ratio on the total number of edges), because it may mislead the partitioner by forcing a strong bias on timing while have no considerations on routability. The parameter set α, β, M, p is empirically obtain through experiments on a benchmark suite. For example, $\alpha, \beta, M, p = 0, 0.25, 6, 20\%$ is reported by [7] based on the best practice considering the MCNC-20 benchmark suite [20].

Graph Partitioning: Circuit partitioning results are achieved by running a k-way graph partitioner, e.g., hMetis [18, 21]. Even though these modern graph partitioners have been well optimized to produce high-quality results in a reasonable runtime, it may not handle high-fanout nets well. In practice, some high-fanout nets, e.g., fanout > 100 is considered in PPack, are removed in hypergraph without affecting routability. To further reduce runtime and manipulate partitioners, a recursive approach is proposed to apply bipartitioning recursively [8]. We refer interested readers to [21] for details about graph partitioning techniques.

Partition refinement: In this step, partitions are refined to ensure they fit FPGA architecture w.r.t. design constraints, such as logic capacity, routability. For example, some partitions from the k-way partitioner may contain more LUTs or FFs than the number of logic resources in a CLB, since these constraints are difficult to force on partitioners. Refinement is applied by incrementally updating partitions. Firstly, all the nodes in the oversized partitions are identified. For each node, the best target partition is determined by computing the gains of merging the node to candidate partitions. The gain function varies from one packer to another. For example, one can use the attraction function as seed-based packers [17]. Once the best target partition is found, the node is moved and the candidate node pool is updated. Such strategy repeats until there are no oversized partitions. To avoid high runtime when there are a large number of candidate partitions, the refinement may be limited to a small range of partitioners, which are considered to be close in hierarchy [8].

Clustered netlist conversion: Refined partitions are converted to clusters one by one. For example, nodes representing BLEs are recoverd to LUTs and FFs, while hidden connections are restored. This is a straightforward process as each partition is already legalized during refinement.

10.1.4 Summary and Trends

Packing has been intensively studied in past decades with several mainstream algorithms proposed. All these algorithms share the same objective: to properly group logic primitives into clusters, so that the number of external nets are minimized and critical path delay is reduced. However, to evaluate the quality of results, packing algorithms cannot be simply judged by the number of external nets and a predicted critical path after packing. Results of packing algorithms are evaluated through a complete design flow, i.e., placement and routing followed by an accurate timing analysis.

Seed-based packers are easy to implement due to their open-source availability and simple nature. Seed-based packers have been extended to support heterogeneous blocks, such as DSPs and BRAMs, while partition-based packers are currently limited to LUT and FFs only. Partition-based packers outperforms seed-based packers on MCNC benchmarks on routability (37% smaller in minimum routable channel width) and timing (12% smaller in critical path delays). But partition-based packers are 10× slower than seed-based packers even for small benchmarks, e.g., MCNC, whose number of nodes are no more than 8 k. High-quality results of partition-based packers can reduce the runtime of placement and routing by 34% on average. However, modern FPGA designs may contain millions of LUTs, which still challenges partition-based algorithms due to their high runtime cost. Choice of packing algorithms also strongly depend on the considered FPGA architectures. Seed-based packers can be tuned through pre-packing and cost functions to fit some dedicated FPGA architectures [22, 23]. Partition-based packers are suitable for FPGA architectures without input bandwidth on clusters, which can efficiently reduce the workloads when rebalancing partitions.

In future, packing algorithms may combine the advantages of seed-based and partition-based (see Table 10.1), while avoid the local optimal and long runtime. For example, the HDPack investigates the use of partitioners for a coarse clustering, and then use seed-based algorithms to perform fast and exact clustering.

10.2 Placement

10.2.1 Overview

FPGA placement is a vital process that having a synthesized design netlist (in terms of clusters/molecules/atoms) to be assigned into physical locations under design constraints (Fig. 10.10).

Good placement is extremely important, because even if the circuit is legally implemented, a poor placement could still lead to a low maximum operating speed or a high power consumption (Fig. 10.11).

Problem Definition: Finding a good placement solution is challenging. A modern commercial FPGA can contain over 2,000,000 function units, exhaustive evaluation is therefore impossible. Besides, not every placement solution is legal, it must under some constraints:

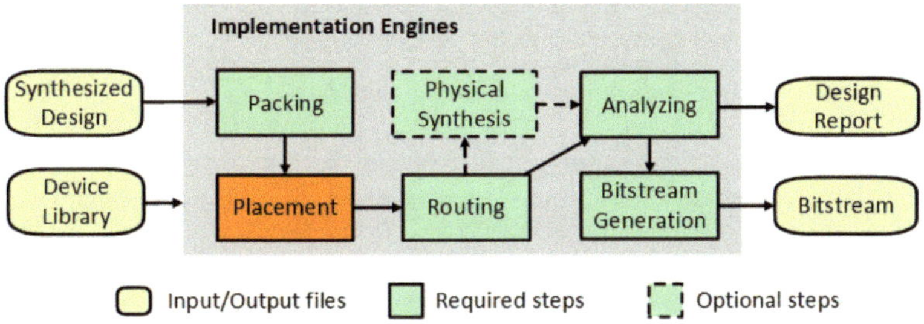

Fig. 10.10 Placement's position in FPGA application EDA flow

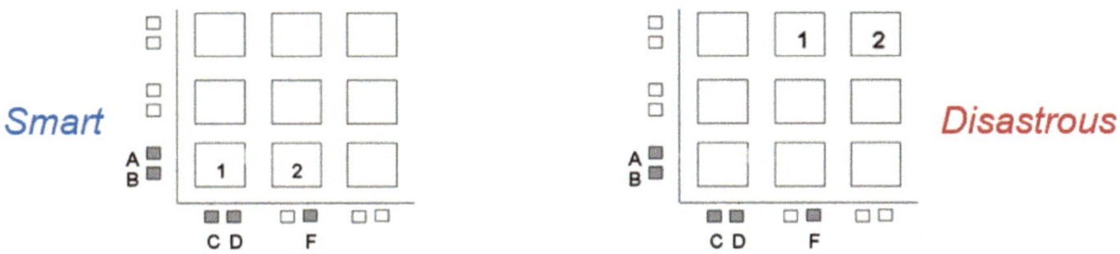

Fig. 10.11 FPGA placement matters

1. Accommodation legality limit
   Generic logic cluster must be placed in a generic logic tile (GLT) location, and a memory cluster must be placed in a memory tile (MMT) location, etc. Placement for FPGAs is actually a slot assignment problem.
2. Specific group limit
   Such as arithmetic logic units forming a carry chain must be placed adjacent to each other in the sequence required by the carry structure.
3. Routing congestion limit
   The circumstance that the interconnect exceeds the fabricated wiring capacity in some part of the FPGA must be avoided.

Mainstream Algorithms: Placement engines can be categorized from different perspectives:

1. By algorithmic logic
   Partition-based techniques [24–26] use divide-and-conquer techniques to recursively partition a circuit and induce ever-smaller placement problems. These techniques can achieve a short runtime, but they suffer from low QoR with large-scale designs and therefore hardly adopted by industry. Annealing and analytical techniques, which will be further discussed in the following sections, are the most popular engines both for academia and industry. In fact, modern placement solutions usually apply them in combination to achieve better results. For example, Intel Quartus placer [27] applies analytical method to determine the an initial placement and then uses annealing method to fine tune it.

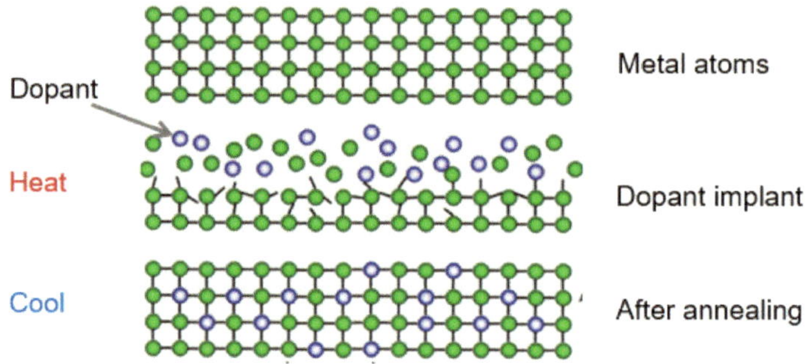

Fig. 10.12 Thermodynamics of physical annealing

2. By optimization objectives
   There are plenty of optimization objectives for placement, for example, wirelength-driven placement minimizes the required wiring; routability-driven placement trying to avoid congestion before routing; timing-driven placement maximizes the circuit speed; power-aware placement takes power consumption into consideration, etc. An excellent placement engine would always balance well among these objectives.
3. By computation acceleration
   In order to reduce compilation time while maintaining quality of results, placement computations can be accelerated either by parallelism or by AI.
4. By targeted architecture
   For other special modern FPGA architecture features (such as advanced package (2.5D/3D), complexed clocking), there are additional placement considerations to adapt to them.

10.2.2 Annealing Placement Algorithms

Physical annealing is a heat treatment that alters the physical and sometimes chemical properties of a material to increase its ductility and reduce its hardness, making it more workable. It occurs by the diffusion of atoms within a solid material, so that the material progresses toward its equilibrium state (Fig. 10.12).

Physical annealing for metal procedures (Fig. 10.13):

Stage1: Heating

Take a metal and heat it to a high temperature. Give it an initial temperature, and atoms transit to high energy states.

Stage2: Cooling

Allow it to cool slowly, metal is annealed to a low temperature. Atoms slowly move to low energy states during the temperature drop.

Higher the initial temperature, slower the cooling, the tougher the metal becomes (Fig. 10.14).

Fig. 10.13 Operation of physical annealing

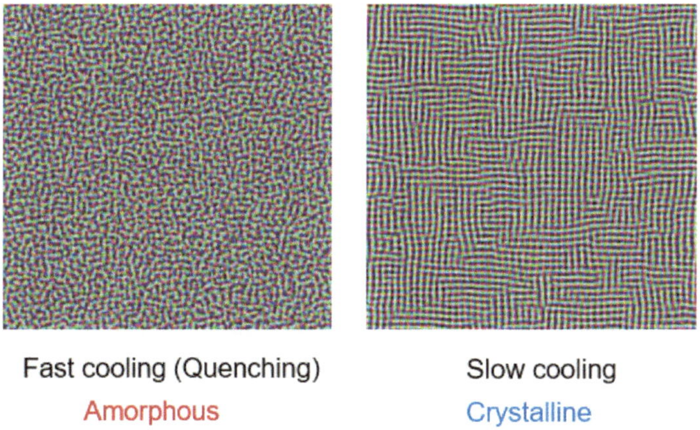

Fig. 10.14 Cooling speed is important in physical annealing

In the field of FPGA EDA, simulated annealing is a probabilistic technique that mimics the physical annealing process, for approximating the global optimum of a given function in a large search space. It was first introduced in 1953 by Metropolis et al. [28] and for solving combinatorial minimization problems and NP complete problem.

Simulated annealing for FPGA placement procedures are (Fig. 10.15):

Stage1: Starts with an initial state and temperature (analogy from thermodynamics).

Stage 2: Randomly changing the state, create a new state.

Stage 3: Compare energies of the new state and the current state, if the new state has less energy, or its probability function $e^{-\Delta C/T}$ is less than a random value (ΔC refers to energy change and T refers to the current temperature), Accept the new state; for circumstances not above, Reject the new state.

```
P = InitialPlacement ();
T = InitialTemperature ();
while (ExitCriterion () == False) do
    while (InnerLoopCriterion () == False) do
        P_new = PerturbPlacementViaMove (P);
        ΔC = Cost(P_new) - Cost(P);
        r = random (0,1);
        if (r < e^(-ΔC/T)) then
            P = P_new;
        end
    end
    T = UpdateTemp (T);
end
```

Algorithm 5: Generic simulated annealing placement algorithm (pseudo-code)

The criteria of simulated annealing placement for FPGA application design includes:

1. Placement Schedule (Annealing)—Defines how to explore the solution space, for example, the starting and ending temperature (the temperature controls the likelihood of accepting moves that make the solution worse); moving strategy; the rate at which the temperature is decreased; the exit criterion for terminating the anneal; the number of moves attempted at each temperature; the method by which potential moves are generated, etc.
2. Cost Function—Energy status of a certain temperature. It is used to evaluate the impact of each proposed move based on the desired optimization objectives.
3. Computation Time—Placement has always been a time-consuming stage, especially for complex FPGA architectures. Research work [29] shows that commercial simulated annealing placer in Quartus II takes about 49% of the compilation time for Titan benchmarks. Reducing computation time is also being intensively studied by both academia and industry.

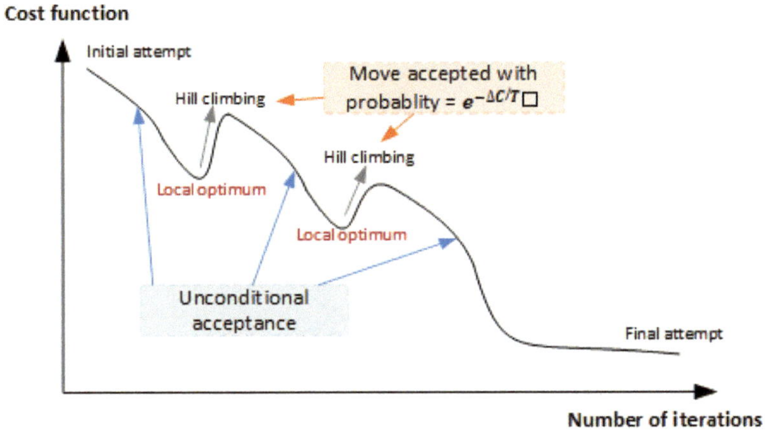

Fig. 10.15 Iteration process of simulated annealing

Placement Schedule (Annealing)

VTR's original VPlace [1] is a famous pioneer work of FPGA placement, according to the algorithm, all the criteria are parameterized (therefore adaptive). In every update, VTR optimize the annealing schedule on the basis of previous version. For example in the newer VTR8 [30], move region limit is optimized by compressed move grid to avoid the situation that preventing a sparse block from moving between columns.

Moving strategy is a researching hot spot in this field. Instead of random move in the original VTR, direct move calculates the effectiveness of each move and make future moves according to this effectiveness [31, 32]. Intel Quartus placer also uses directed moves, but no details has been published [33].

Cost Function

1. Wirelength
   Wirelength, often measured in Half Perimeter WireLength (HPWL), is the most important objective that can not be neglected.
   Bounding box (the smallest rectangle to encloses all the terminals of a net) is introduced to estimate the wiring cost (Fig. 10.16) and the cost function (HPWL of the bounding box) is defined as (Eq. 10.3). bb_x and bb_y are the x- and y-directed span of the bounding box that just encloses all the terminals of the net (Fig. 10.16). With this cost function, place engine could minimize the total HPWL across all considered nets.

$$\text{WireCost} = \sum_{i=1}^{\text{num\_nets}} [bb_x(i) + bb_y(i)] \tag{10.3}$$

   VPR's VPlace [1] uses linear congestion cost function to take routability into consideration (Eq. 10.4).

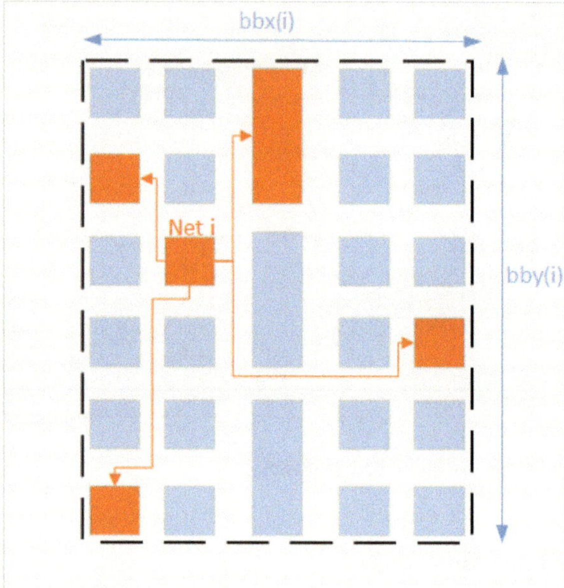

Fig. 10.16 Bounding box is used to estimate the wiring cost

$$\text{WireCost} = \sum_{i=1}^{\text{num\_nets}} q(i) \left[\frac{bb_x(i)}{[C_{\text{av},x}(i)]^\beta} + \frac{bb_y(i)}{[C_{\text{av},y}(i)]^\beta} \right] \tag{10.4}$$

where $q(i)$ is fanout-based correction factor and linearly increases to help correct the underestimated wiring. $C_{\text{av},x}(i)$ and $C_{\text{av},y}(i)$ are the average channel capacities (in tracks) in the x and y directions respectively, over the bounding box of net i. β allows the relative cost of using narrow and wide channels to be adjusted. The larger value of β, the more wiring in narrow channels is penalized relative to wiring in wider channels.

2. Timing

   The timing cost function can be loosely classified as net-based [34–38], path-based [39, 40] or a hybrid of the two [41, 42]. Net-based cost function usually transform timing to net weights, while the path-based cost function try to represent the timing of critical paths directly. Using path-based cost function generally has more accurate timing view and control, but it suffers from poor scalability and high complexity; using the net-based cost function is very suitable for large chip design such as FPGAs since it has relatively low computational complexity and high flexibility.

   T-VPlace [36] is the timing-driven version of VPlace, it improves the cost function by adding net-based timing considerations (Eq. 10.5).

$$\text{TimingCost} = \sum_{j=1}^{\text{num\_nets}} \text{Criticality}(j) \cdot \text{Delay}(j) \tag{10.5}$$

where Delay(j) is the delay value of the edge joining node j to node i. For quick estimation, it computes the delay between two blocks as a function only of the distance (δx, δy). Before that, the placer employs router to determine the delay between two blocks that are (δx, δy) distance apart, and record it in the delay look up table at location index (δx, δy). Criticality(j) (Eq. 10.6) of connection j in the design is determined by periodic timing analysis.

$$\text{Criticality}(j) = 1 - \frac{\text{Slack}(j)}{\text{Delay}_{\text{max}}} \tag{10.6}$$

where $\text{Delay}_{\text{max}}$ is the critical path delay (maximum arrival time of all sinks in the circuit), and Slack(j) is the amount of delay that can be added to the connection j without increasing the critical path delay.

3. Power
   The power cost function is dependent on the switching activity of the hardware resources of FPGA.
   Lamoureaux [43] modified VPR's cost function by adding a power cost (Eq. 10.7):

$$\text{PowerCost} = \sum_{i=1}^{\text{num\_nets}} q(i)[bb_x(i) + bb_y(i)] \cdot \text{activity}(i) \tag{10.7}$$

   where activity (i) represents the average number of times net i transitions per second.

Total cost function often trades off these objectives above. If the trade-off variable λ determines how much weight to give timing cost, variable γ determines how much weight to give power cost, then the total cost could be described as (10.8).

$$\text{WeightedCost} = \lambda\text{TimingCost} + \gamma\text{PowerCost} + (1 - \lambda - \gamma)\text{WiringCost} \tag{10.8}$$

Computation Time

1. Parallel acceleration
   Distribute placement moves among multiple computation engines to be evaluated in parallel is widely used to shorten annealing time. However, this could lead to conflicts if multiple computational engines accept moves that affect the same design units or nets (termed as "collision") and nondeterministic or serial nonequivalent results.
   Parallel solutions address the problems above and can be inspected from different perspective:

a. In terms of algorithmic logic (Software)
Independent set identifying—Find independent (non-colliding) set of moves and process them all in parallel. The Intel Quartus placer [33] is a representative work in this field, it speculatively evaluates moves in parallel and uses a dependency checker to detect collisions.
Relevant set partitioning—Assign each computation core a partition in the placement area such that different computation unit's moves will not interfere with each other. An early parallel implementation for standard cell placement [44] uses this method to avoid collision.
b. In terms of computation architecture (Hardware)
Scalar architecture (CPU)—Modern CPU usually has multiple cores, and each core has numbers of threads. Using multiple CPU cores or threads as computation engines for annealing placement has been intensively studied [33, 45–48].
Vector architecture (GPU)—Streaming multiprocessor (SMP) is the computation engine of GPU and can be used as annealing placement computation engine to obtain deterministic result. Related works [49, 50] have achieved decent speedups over multi-threaded CPU implementations.
Spatial architecture (FPGA)—In [51], systolic array of processing element (PE) in an FPGA is used as the computation engine. It accelerates the annealing placer and achieves a speedup of up to 2649× over VPR run with the fast option, at a cost of 36% average increase in minimum channel width for successful routing. However, the key limitation of this work is requiring an FPGA that is at least 400 times larger than the circuit being placed, making it unusable for large designs.

2. AI acceleration
AI technologies has been emerging in the past years to guide the annealing placer in choosing which type of move to make and greatly reduce computation time.

a. In terms of algorithmic logic (Software)
Reinforcement learning (RL)—is a branch of machine learning, it utilizes a software agent to make observations and takes actions within an environment, and its objective is to learn to act in a way that will maximize its expected long-term rewards. Consider the task of FPGA annealing placement [32, 52], the goal is to teach the annealer (agent) to make move decisions with reinforcement learning, given the current placement status (environment) S_t, RL techniques iteratively use an action value function $Q(s, a)$ to predict the immediate and future expected cost optimization(reward) if action a is chosen while the environment is in state s. Any action chosen will not only affect the immediate reward but also all the upcoming rewards and states. After performing action a_t and receiving reward r_{t+1}, the action value function Q can be updated as (Eq. 10.9):

$$Q(a_{t+1}) = Q(a_t) + \lambda(r_{t+1} - Q(a_t)) \tag{10.9}$$

where $r_{t+1} - Q(a_t)$ represents the deviation between the agent's estimate and the actual reward, and λ is the step size parameter to reduce this deviation.

b. In terms of computation architecture (Hardware)
Scalar architecture (CPU)—Known studies deploy the agent on CPUs, such as [32, 52].
In fact, competitive rivals such as vector (GPU) or matrix (TPU) engines could be even more efficient for this type of workload. However, none of these works is published by the time this book is written.

10.2.3 Analytical Placement Algorithms

Analytical placement methods consider global connectivity rather than evaluate the local modifications, however, the global minimum is usually an illegal placement with overlapping blocks, so constraints and heuristics must be applied to guide the algorithm to a legal solution. At the 2016 International Symposium on Physical Design (ISPD) contest, analytical placers occupied the top three positions (UTPlace [53], Ripple [54] and GPlace [55]).

Most modern FPGA analytical placers consist of the following three major actions:

Action 1: Global placement (GP). This action ignores some constraints (e.g., unit overlaps) and computes the best position (coordination) for each unit according to desired objectives (e.g., wirelength). It has a crucial impact on the overall placement quality.

Action 2: Legalization (LG). This action removes all overlaps among design units, assigning each of them into device units.

Action 3: Detailed placement (DP). This action further improves the legalized placement solution, typically in an iterative manner by rearranging a small number of units in a given region while keeping all other units fixed.

Among these actions, GP is the most time-consuming one. As a reference, elfPlace [56] reports that in terms of runtime breakdown, GP, LG, and DP consumes 59.8%, 19.9%, and 18.6%, respectively.

The criteria of analytical placement for FPGA application design includes:

1. Placement Schedule (Analytical)—Defines how to explore the solution space, or the algorithmic strategies of invoking each action.
2. Cost Function—It is used to evaluate the impact of each proposed status change based on the desired optimization objectives.
3. Computation Time—Having the same concerns with annealing techniques in the previous section, reducing placement time is an endless optimization direction for every researcher.

Placement Schedule (Analytical)

There are several different algorithmic concerns with FPGA analytical placement.

For GP, analytical solving process tends to pull FPGA units together for better PPA, so that the placement strategy is highly depends on the cost function, which will be discussed later in this section.

For LG, it is assumed that different PPA objectives are already optimized in the GP. Consequently, to preserve the GP result as far as possible, the objective of LG is to minimize the movement of the FPGA units. Partition-based methods are widely used due to their simplicity [54, 57, 58], during spreading, the FPGA will be evenly divided into a grid of x*y bins, and the legalizer will find an overflowed bin, expand it into a corresponding larger window, recursively partition the window and spread the units within it.

For DP, the objective is to find a better position for each FPGA unit in the available free spaces. These detailed refinements are often performed using low temperature simulated annealing to fully optimize the FPGA design [54, 59].

Cost Function

1. Wirelength
   Wirelength cost functions have two types of models to approximate HPWL: quadratic and nonlinear.
   Quadratic cost functions are mostly used for analytical placement [54, 57, 59], and the wirelength is modeled in quadratic formula:

$$\text{WireCost}_{\text{quadratic}} = \frac{1}{2}\sum_{i,j} w_{i,j}[(x_i - x_j)^2 + (y_i - y_j)^2] \tag{10.10}$$

   where $w_{i,j}$ is the weight of the connection between FPGA unit i and j. x and y denote the unit locations. The cost function can be written in matrix notation as:

$$\text{WireCost}_{\text{quadratic}} = \frac{1}{2}x^T Q_x x + c_x^T x + \frac{1}{2}y^T Q_y y + c_y^T y + \text{Constant} \tag{10.11}$$

   For the x dimension, a matrix, Q_x, represents connections between movable units (i.e., units being placed), and a vector, c_x, represents connections between movable and fixed units.
   The cost function can be further separated into x and y components and it is equivalent to solving:

$$Q_x + c_x = 0;\ Q_y + c_y = 0 \tag{10.12}$$

   Once formula (Eq. 10.12) is solved, the x and y locations of the FPGA units is settled.
   Nonlinear cost functions, on the other hand, using higher-order formula to represent wirelength. They formulate the nonoverlap constraint using differentiable nonlinear functions and solve it together with the wirelength in a unified objective function. Using nonlinear cost functions can often get higher solution quality, however, this quality improvement also comes with longer runtime due to the more expensive nonlinear optimization. Logarithm-sum-exponent (LSE) is a typical nonlinear expression to approximate the HPWL (Eq. 10.13).

$$\text{WireCost}_{\text{lse}} = \gamma(\log \sum_{v_i \in e} e^{\frac{x_i}{\gamma}} + \log \sum_{v_i \in e} e^{\frac{-x_i}{\gamma}}) \tag{10.13}$$

where e is the target net and $v_i \in e$ is the units in that net. When γ is equal to zero, the LSE model is reduced to the exact HPWL. However, in practical implementation, a small reasonable γ value is chosen to avoid arithmetic overflow.

2. Timing
   See timing cost of annealing methods.
3. Power
   See power cost of annealing methods.

Just like simulated annealing methods, total cost function will trades off these objectives above.

Computation Time

1. Parallel acceleration
   Since detailed placement can be seen as a low temperature annealing process, which we have discussed in the previous section, global placement and legalization can also benefit from parallel acceleration techniques:

   a. In terms of algorithmic logic (Software)
      Analytical solver for the x and y dimensions can be assigned to different computation unit. This parallelization could result in close to $2\times$ improvement in computation time (with 2 CPU threads) [57].
   b. In terms of computation architecture (Hardware)
      Scalar architecture (CPU)—Using multiple CPU cores or threads as computation engines for analytical placement is the most ordinary way [57, 60].
      Vector architecture (GPU)—Nonlinear placement (using nonlinear cost function) usually performs better than quadratic placement (using quadratic cost function), although the difference in quality is small [61]. Acceleration of nonlinear analytical placement on GPUs then emerged [62, 63].
      Spatial architecture (FPGA)—*In the first work on acceleration of analytical placement on FPGAs [64], wirelength gradient is computed by using OpenCL.

2. AI acceleration
   AI technologies has been emerging in the past years to aid FPGA analytical placers to reduce computation time.

   a. In terms of algorithmic logic (Software)
      Deep learning (DL)—is a branch of machine learning, it uses data to train a model to make predictions from new data. For FPGA analytical placement, the first attempt [65] uses deep-learning to direct the placer's optimization strategy. In this work, a Convolutional Encoder-Decoder (CED) is utilized to predict the congestion present in subsequent placement iterations including the final placement, and achieve reductions in placer runtime between 27 and 40% with no significant deterioration in quality-of-result. [66] presents a CNN-inspired analytical placement algorithm to effectively handle the redundant frequency

translation problem for large-scale FPGAs.
Reinforcement learning (RL)—Discussed in previous section, RL framework can be integrated into detailed placement. In [67], several RL models are developed to capture the different characteristics of placement solutions and use them to guide decision making process. As a result, 50% of the CPU time is saved on the basis of [59].

b. In terms of computation architecture (Hardware)
Scalar architecture (CPU)—CPUs are still the most traditional computation platform for AI-aided FPGA analytical placement [65].
Works on vector (GPU) or matrix (TPU) computing platforms still needs time to emerge.

10.2.4 Summary and Trends

As the two major approaches for FPGA placement, annealing, and analytical engines have their own strengths and weaknesses.

1. For annealing
It has better quality for small designs, and easier to consider multiple objectives simultaneously, however, it is slower for large circuits, and sometimes could meet freezing problem—unable to escape local minima.
2. For analytical
The multi-stage (GP-LG-DP) characteristics make it a de facto hybrid placement technique that is highly flexible. It is more efficient and scalable for large designs with considerably good quality, however it still face its own challenges, such as targeting mix-size FPGA units, integrating timing/power metrics into the optimization objective. [68].
FPGA placement definitely will face new challenges continuously as FPGA architecture evolves. First, multiple objectives need to be considered at the same time due to the increasing complexity of modern FPGAs. It's usually difficult to make a trade-off among multiple objectives. Thus, an optimal resolution that performs good in every aspect will become harder to achieve. Secondly, extra architectural constraints are often extremely difficult to handle in placement due to their discreteness and irregularity. Last but not least, how to save computation time and memory footprint, especially when the FPGA design is gargantuan, still haunted every practitioners in this field.

10.3 Routing

10.3.1 Overview

FPGA routing is a process that determines which programmable switches should be turned on to connect all the logic unit input and output pins required by the circuit (Fig. 10.17).

Logic units in the application design must be well connected by using device's interconnection resources to actually make the circuit work, because a poor routing engine will lead to a lower maximum operating speed, greater power consumption, slower implementation time or even a failure to route all signals. The constraints is relatively simple: two nets cannot be routed on the same wire, which may cause routing congestion.

FPGA is like a city on a grid, routing is like solving the traffic problems in the city. Two fundamental steps must be carried out—building the traffic infrastructure (building routing resource graph under the help of routing guidance model) and navigating each trip (route all signals with routing engines).

Problem Definition: For each application design net, source is at which all nets begin, sink is at which all net terminals end. There can be one source (s_i) and multiple sinks ($T\_i = t\_i1, t\imath 2, \ldots t\_ik$, which refers to all k sinks of s_i), so the application design has set of sources ($S = s\_1, s\_2, \ldots s\_i$) and set of sets of sinks ($T = T\_1, T\_2, \ldots T\_i, T\_i = t\_i1, t\imath 2, \ldots t\_ik$). Routing problem is to find paths from each source s_i to all sinks in T_i, paths emanating from different sinks that must be disjoint (cannot share any nodes or edges).

Similar to placement, routing is extremely important for FPGA application design, since a poor solution will lead to a lower maximum operating speed, increase power consumption, slow implementation time or even a failure to route all signals. Given the fact that routing is a NP complete problem, finding a good routing is challenging, let alone relative scarcity of routing resources and signals will compete for the same routing resources.

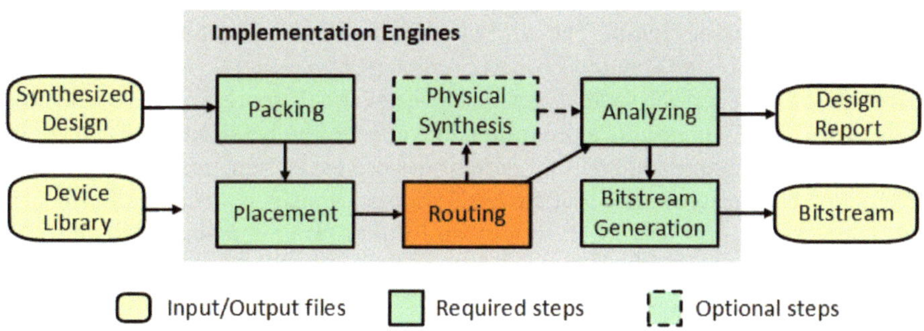

Fig. 10.17 Routing's position in FPGA application EDA flow

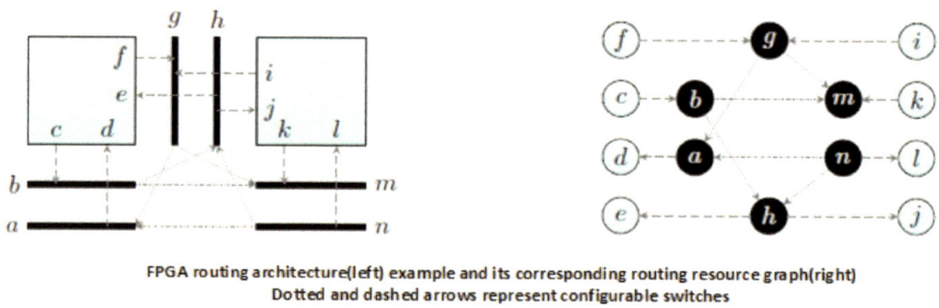

FPGA routing architecture(left) example and its corresponding routing resource graph(right)
Dotted and dashed arrows represent configurable switches

Fig. 10.18 FPGA RRG example [69]

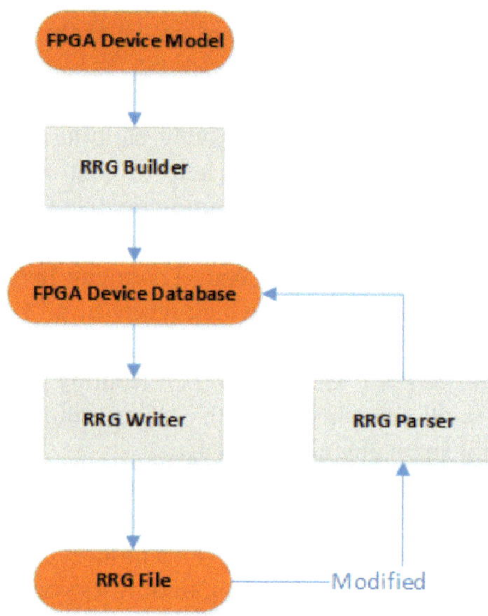

Fig. 10.19 FPGA RRG processing flow

Routing Resource Graph: The primary data structure representing FPGA routing resources is the directed Routing Resource Graph (RRG) $G = (V, E)$, where V is the set of vertices and E is the set of edges. Each vertex $v \in V$ represents wires and pins, each edge $e_{ij} \in E$ represents a programmable connection between a pin and a wire segment, or a programmable routing switch between two wire segments (Fig. 10.18). Programmable switches can be fabricated as pass transistors, tri-state buffers, or multiplexers. Multiplexers are the dominant form of programmable interconnect in recent FPGAs due to a superior area-delay product and thus unidirectional.

RRG can be represented by readable file. VTR-XML [70] is the only publicly known format to describe RRG. The RRG building, parsing and writing flow can be described as (Fig. 10.19).

Mainstream Algorithms: Routing engines can be categorized from different perspectives:

1. By algorithmic logic
   Negotiation-based routing algorithms[69, 71] typically consist of two major steps: path searching and congestion removal. During path searching, it explore all feasible paths for each net and identify the best one. After path searching, rip-up and rerouting operation is carried out to eliminate congestion. Boolean-based routing algorithms [72–76] transform the FPGA routing task to one of simultaneously satisfying a set of Boolean constraints. If successful, the solution found by the solver is converted into a valid route, otherwise the signal is unroutable.
   The negotiation-based routing algorithm is in dominant use in the FPGA community due to its superior performance and quality of results.
2. By optimization objectives
   Routability-driven routing prioritizes congestion avoidance as its primary goal[77]; timing-driven routing maximizes the circuit speed[69, 78, 79]; power-aware routing saves the power consumption [43, 80], etc. An excellent routing engine would always balance well among these objectives.
3. By computation acceleration
   In order to reduce compilation time while maintaining quality of results, routing computations can be accelerated either by parallelism or by AI.
4. By targeted architecture
   For other special modern FPGA architecture features (such as advanced package (2.5D/3D), complexed clocking), there are additional routing considerations to adapt to them.

10.3.2 Negotiation-Based Routing Algorithms

Negotiation-based routing methods route nets one by one. Larry McMurchie and Carl Ebling proposed a negotiation approach in 1995, called Pathfinder [71], which is very famous in routing society. Pathfinder first gives every edge (connection) in RRG a cost that depends on current usage and historical usage, then each net is routed by a breath first searching (BFS), making the cost the lowest. Multiple nets may use the same node in RRG (flagged as a congestion). A congestion node is given a higher cost. If a route must include a congested node, it will "negotiate" with the other routes and make them go around (rip-up and re-route).

The criteria of negotiation-based routing for FPGA application design includes:

1. Routing Schedule (Negotiation-based)—Defines how to explore the solution space, or the algorithmic strategies of invoking each action.
2. Cost Function—It is used to evaluate the impact of each proposed status change based on the desired optimization objectives.

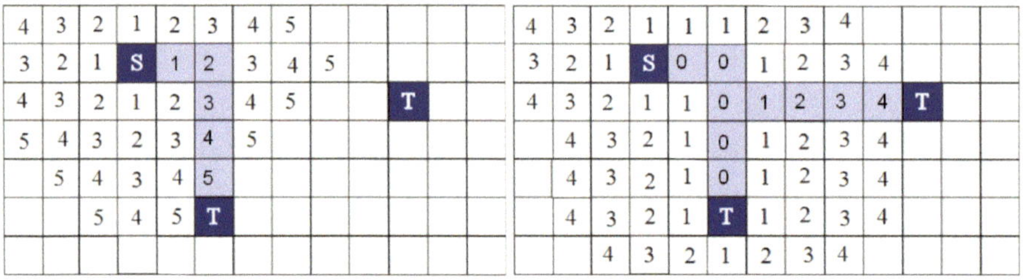

(a) Search the path from the source to the first sink (b) Set the cost of nodes on the previous path to 0 and search for the second sink

Fig. 10.20 Routing schedule for multiple sinks in the same net

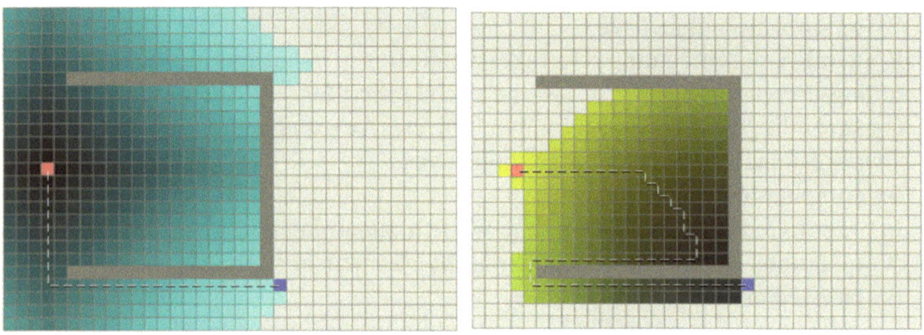

(a) BFS without direction (b) BFS with direction

Fig. 10.21 When BFS meets a concave obstacle

3. Computation Time—Having the same concerns with annealing techniques in the previous section, reducing placement time is an endless optimization direction for every researcher.

Routing Schedule (Negotiation-based)
Route nets in decreasing order of fanout is a common schedule, because high-fanout nets tend to span the larger area of FPGA and easier to route when there is less congestion to other nets routed earlier, low fanout nets tend to be more localized and relatively easier to route even in the presence of some congestion.

When routing a net with one source and multiple sinks, breath first searching (BFS) is invoked to find the first sink like the process of wave expansion. The wavefront always have the biggest cost. After that, the cost of nodes on the previous path is set to 0 and the second sink is searched iteratively (Fig. 10.20).

In general, BFS without direction is time-consuming and directed search is far more effective, however, when the obstacle is concave, BFS without direction finds the best route but time-consuming, BFS with direction is time saving but sometimes can not find the best route (Fig. 10.21).

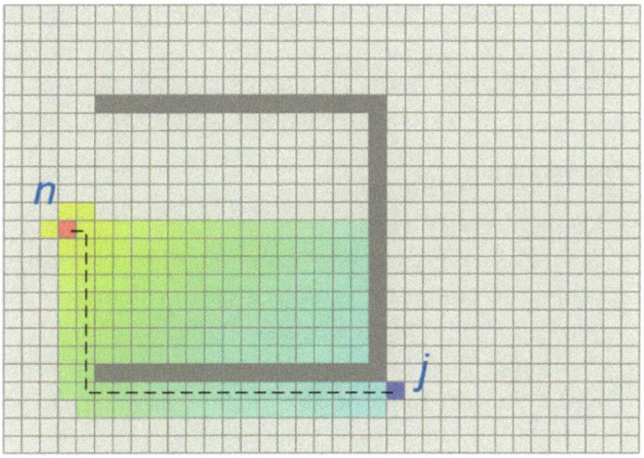

Fig. 10.22 Banlanced BFS searching in FPGA routing

Therefore, cost function must be proper balanced between directed searching and directionless searching. Just like authoritarian and democracy, a good political governance model would always balance them well (Fig. 10.22).

The general negotiation process's pseudo code is shown in Algorithm 6.

```
Let: RT_i be the set of nodes in the current routing of net i;
while shared resources exist(routing congestion occurs) do
    foreach net, i do
        rip-up routing tree RT_i;
        RT_i = s_i;
        foreach sinkt_ij do
            Initialize priority queue PQ to RT_i at cost 0;
            while sink t_ij not found do
                Remove lowest cost node m from PQ;
                foreach fan-out node n of node m do
                    Add n to PQ at PathCost(n) = Costn + PathCost(m);
                end
            end
            foreach node n in path t_ij to s_i do
                Update c_n;
                Add n to RT_i;
            end
        end
    end
    update historical cost for n;
end
```

Algorithm 6: Negotiation process in negotiation-based routing algorithm (pseudo-code)

In VTR routing, this consideration is expressed by adding an ExpectedCost(n, j) to the PathCost(n) (Eq. 10.14).

$$\text{TotalCost}(n) = \text{PathCost}(n) + \alpha \cdot \text{ExpectedCost}(n, j) \tag{10.14}$$

where ExpectedCost(n, j) is an estimated cost from current node n to target sink j, PathCost(n) is the cost of the path from the current partial routing tree to node n.

Cost Function

1. Routability (Congestion)
   In real life, there has been a system of surcharging for use of public roads that are subject to congestion through excess demand. This traffic control measure has been applied to many big cities (Fig. 10.23).
   Pathfinder [71] has set a classic example of routability cost, for node n in the RRG, the cost function is presented as (Eq. 10.15).

$$\text{RoutabilityCost}(n) = (b(n) + h(n)) \cdot p(n) \tag{10.15}$$

   where $b(n)$ is the base cost of a routing through node n, $h(n)$ is related to the history of congestion on node n during previous iterations, and $p(n)$ is related to the number of nets (signals) presently routed through node n at the current iteration.
   VTR router [69] is based upon the Pathfinder negotiated congestion algorithm, the modified routability cost it defined is shown as (Eq. 10.16).

$$\text{RoutabilityCost}(n) = b(n) \cdot h(n) \cdot p(n) + \text{Bend}(n, m) \tag{10.16}$$

   where $b(n)$, $h(n)$ and $p(n)$ has exactly the same meaning with those in the Pathfinder, the additional parameter Bend (n, m) penalize global routes with bends since these routes are less likely to use long wires, making detailed routes difficult to implement (congestion more likely to occur). Meanwhile, multiplying $b(n)$ and $h(n)$ eliminates normalization.
2. Timing
   Cost of using a RRG node n in the routing of a connection c can be represented as (Eq. 10.17).

$$\text{TimingCost}(n) = \text{TimingCriticality}(c) \cdot \text{Delay}(n) \tag{10.17}$$

   The timing criticality is the ratio of the connection slack to the longest delay in the circuit (Eq. 10.18).

$$\text{TimingCriticality}(c) = 1 - \frac{\text{Slack}(c)}{\text{Delay}_{\text{max}}} \tag{10.18}$$

   where $\text{Delay}_{\text{max}}$ is the critical path delay (maximum arrival time of all sinks in the circuit), and Slack(c) is the amount of delay that can be added to the connection c without increasing the critical path delay.
   In Pathfinder, it uses TimingCriticality(c) as the weighted factor of timing cost.

3. Power
   Lamoureaux [43] modified VPR's cost function by adding a power cost (Eq. 10.19):

$$\text{PowerCost}(n) = \text{ActCriticality}(i) \cdot \text{Cap}(n) \tag{10.19}$$

   where $\text{Cap}(n)$ is the capacitance associated with routing resource node n and $\text{ActCriticality}(i)$ is the activity criticality of net i.
   In [43], it uses $\text{ActCriticality}(i)$ as the weighted factor of timing cost.

$$\text{ActCriticality}(i) = \min\left(\frac{\text{Act}(i)}{\text{MaxAct}}, \text{MaxActCriticality}\right) \tag{10.20}$$

   where $\text{Act}(i)$ is the switching activity in net i, MaxAct is the maximum switching activity of all the nets, and MaxActCriticality is the maximum activity criticality that any net can have.

Total cost function often trades off these objectives above. If the trade-off variable λ determines how much weight to give timing cost, variable γ determines how much weight to give power cost, then the total cost could be described as (Eq. 10.21).

$$\text{WeightedCost} = \lambda\text{TimingCost} + \gamma\text{PowerCost} + (1 - \lambda - \gamma)\text{RoutabilityCost} \tag{10.21}$$

Computation Time

1. Parallel acceleration
   Researchers have been tirelessly looking for ways to accelerate FPGA routing through parallelism, since routing is one of the most time-consuming compilation step in the whole flow. Most of the parallel acceleration work for FPGA routing focuses on the negotiation-based methods. An ideal parallel router is not only fast, but also scalable and deterministic.

Fig. 10.23 Congestion charging in central London

a. In terms of algorithmic logic (Software)
   Almost all parallel routers are either coarse-grain or fine-grain. Coarse-grain parallel routers [81–83] distribute the problem by partitioning the nets and then route them independently, whilst fine-grain parallel routers accelerate the routing of a single net. For coarse-grain parallel routers, nets are partitioned based on the independence of their bounding boxes. If the bounding boxes of two nets overlap, the possibility of conflicts between the nets is high and they should probably not be routed in parallel. For fine-grain parallel routers [84], works for an individual net such as maze expansion could be accelerated.
b. In terms of computation architecture (Hardware)
   Scalar architecture (CPU)—Most of the parallelization work on FPGA routing is based on CPU [81–83].
   Vector architecture (GPU)—GPU-accelerated EDA has been studied for years [85], however, the first published work on utilizing GPU to accelerate FPGA routing [86] came out later in 2017. It leverage Bellman-Ford algorithm to optimize the computation and experimental results show that an average of 18.72% speedup is achieved.
   Spatial architecture (FPGA)—[87] is the first attempt on FPGA-accelerated FPGA routing, however, it ends up with 4–6× slower than running on pure CPU platform due to the limited performance of the chosen hardware platform (mid-end ARM+FPGA SoC).

2. AI acceleration
   FPGA negotiation-based router start to benefit from AI technologies from the 2010s.

   a. In terms of algorithmic logic (Software)
      Reinforcement learning (RL)—The agent is guided toward achieving legal routing solution by formulating a reward function. Every time an action is taken by an agent while moving from one node to another, it must get a reward that is added to its experience for any future moves regarding that particular node. In [88], the reward formulation is defined as (Eq. 10.22), showing that the objective of the routing is to minimize the number of conflicts.

$$r_t = -\Delta_{\text{conflict}} \tag{10.22}$$

      The agent continuously learns and adjusts the award values until get the optimized solution. It is reported that on average, the RL technique can reduce 30% routing time for similar quality of results.
   b. In terms of computation architecture (Hardware)
      Scalar architecture (CPU)—CPUs are still the most traditional computation platform for AI-aided FPGA negotiation-based routing.
      Works on vector (GPU) or matrix (TPU) computing platforms still needs time to emerge.

10.3.3 Summary and Trends

Modern FPGA routing are essentially RRG searching problem, balancing concerned metrics such as routability, timing, and power. Negotiation-based algorithms have been proved to be the most efficient solvers, and hence have ruled the FPGA routing world for years. VTR is the most popular open-sourced implementation of negotiation-based FPGA router. How to further refine it in terms of QoR or computation time has been the main researching topic and this trend will continue in the foreseeable future.

AI(ML)-aided techniques has emerged in accelerating FPGA routing tasks. These intelligent routing algorithms can significantly improve the efficiency and reliability of FPGA designs, while also reducing the design time and cost [89]. As the study deepens, AI-aided router will play an increasingly important role.

References

1. V. Betz, J. Rose, VPR: a new packing, placement and routing tool for FPGA research. FPL (1997)
2. A.S. Marquardt, V. Betz, J. Rose, Using cluster-based logic blocks and timing-driven packing to improve FPGA speed and density, in *Proceedings of the 1999 ACM/SIGDA Seventh International Symposium on Field Programmable Gate Arrays*, ser. FPGA '99. (Association for Computing Machinery, New York, NY, USA, 1999), pp. 37–46. [Online]. Available: https://doi.org/10.1145/296399.296426
3. E. Bozorgzadeh, S. Ogrenci-Memik, M. Sarrafzadeh, Rpack: Routability-driven packing for cluster-based FPGAs, in *Proceedings of the ASP-DAC 2001. Asia and South Pacific Design Automation Conference 2001 (Cat. No.01EX455)* (2001), pp. 629–634
4. A. Singh M. Marek-Sadowska, Efficient circuit clustering for area and power reduction in FPGAs, in *Proceedings of the 2002 ACM/SIGDA Tenth International Symposium on Field-Programmable Gate Arrays*, ser. FPGA '02. (Association for Computing Machinery, New York, NY, USA, 2002), pp. 59–66. [Online]. Available: https://doi.org/10.1145/503048.503058
5. J. Luu, J. H. Anderson, J.S. Rose, Architecture description and packing for logic blocks with hierarchy, modes and complex interconnect, in *Proceedings of the 19th ACM/SIGDA International Symposium on Field Programmable Gate Arrays*, ser. FPGA '11. (Association for Computing Machinery, New York, NY, USA, 2011), pp. 227–236. [Online]. Available: https://doi.org/10.1145/1950413.1950457
6. J. Luu, J. Rose, J. Anderson, Towards interconnect-adaptive packing for FPGA, in *Proceedings of the 2014 ACM/SIGDA International Symposium on Field-Programmable Gate Arrays*, ser. FPGA '14 (Association for Computing Machinery, New York, NY, USA, 2014), pp. 21–30. [Online]. Available: https://doi.org/10.1145/2554688.2554783
7. W. Feng, K-way partitioning based packing for FPGA logic blocks without input bandwidth constraint, in *2012 International Conference on Field-Programmable Technology* (2012), pp. 8–15
8. W. Feng, J. Greene, K. Vorwerk, V. Pevzner, A. Kundu, Rent's rule based FPGA packing for routability optimization, in *Proceedings of the 2014 ACM/SIGDA International Symposium on Field-Programmable Gate Arrays*, ser. FPGA '14 (Association for Computing Machinery, New York, NY, USA, 2014), pp. 31–34. [Online]. Available: https://doi.org/10.1145/2554688.2554763

9. D.T. Chen, K. Vorwerk, A. Kennings, Improving timing-driven FPGA packing with physical information, in *2007 International Conference on Field Programmable Logic and Applications* (2007), pp. 117–123.
10. Z. Huang, Z. Li, N. Wang, P. Tao, X. Zhou, L. Wang, Repack: a packing algorithm to enhance timing and routability of a circuit, in *2012 IEEE 11th International Conference on Solid-State and Integrated Circuit Technology* (2012), pp. 1–5
11. S.T. Rajavel, A. Akoglu, Mo-pack: many-objective clustering for FPGA CAD, in *2011 48th ACM/EDAC/IEEE Design Automation Conference (DAC)* (2011), pp. 818–823
12. K.E. Murray, O. Petelin, S. Zhong, J.M. Wang, M. Eldafrawy, J.-P. Legault, E. Sha, A.G. Graham, J. Wu, M.J.P. Walker, H. Zeng, P. Patros, J. Luu, K.B. Kent, V. Betz, Vtr 8: high-performance cad and customizable FPGA architecture modelling. ACM Trans. Reconfigurable Technol. Syst. **13**(2) (2020). [Online]. Available: https://doi.org/10.1145/3388617
13. G. Karypis, V. Kumar, Multilevel k-way hypergraph partitioning, in *Proceedings of the 36th Annual ACM/IEEE Design Automation Conference*, ser. DAC '99 (Association for Computing Machinery, New York, NY, USA, 1999), pp. 343–348. [Online]. Available: https://doi.org/10.1145/309847.309954
14. J. Luu, Architecture-aware packing and cad infrastructure for field-programmable gate arrays, Ph.D. dissertation, University of Toronto (2014)
15. K. Keutzer, Dagon: technology binding and local optimization by DAG matching, in *24th ACM/IEEE Design Automation Conference* (1987), pp. 341–347
16. G. Chen, J. Cong, Simultaneous timing driven clustering and placement for FPGAS, in *International Conference on Field Programmable Logic and Applications* (Springer, 2004), pp. 158–167
17. Z. Marrakchi, H. Mrabet, H. Mehrez, Hierarchical FPGA clustering based on multilevel partitioning approach to improve routability and reduce power dissipation, in *2005 International Conference on Reconfigurable Computing and FPGAs (ReConFig'05)* (2005), pp. 4–25
18. G. Karypis, V. Kumar, Hmetis: a hypergraph partitioning package. ACM Trans. Architect. Code Optim. (1998)
19. H. Zhang, L. Song, Z. Han, Y. Zhang, Basics of hypergraph theory, in *Hypergraph Theory in Wireless Communication Networks* (Springer, 2018), pp. 1–19
20. S. Yang, Logic synthesis and optimization benchmarks user guide: version 3.0. Citeseer (1991)
21. S. Schlag, T. Heuer, L. Gottesbüren, Y. Akhremtsev, C. Schulz, P. Sanders, High-quality hypergraph partitioning. ACM J. Exp. Algorithmics (2022). [Online]. Available: https://doi.org/10.1145/3529090
22. X. Tang, P.-E. Gaillardon, G. De Micheli, Pattern-based FPGA logic block and clustering algorithm, in *2014 24th International Conference on Field Programmable Logic and Applications (FPL)* (2014), pp. 1–4
23. P.-E. Gaillardon, X. Tang, G. Kim, G. De Micheli, A novel FPGA architecture based on ultrafine grain reconfigurable logic cells. IEEE Trans. Very Large Scale Integr. (VLSI) Syst. **23**(10), 2187–2197 (2015)
24. P. Maidee, C. Ababei, K. Bazargan, Fast timing-driven partitioning-based placement for island style FPGAs, in *Proceedings 2003. Design Automation Conference (IEEE Cat. No.03CH37451)* (2003), pp. 598–603
25. A. Khatkhate, C. Li, A.R. Agnihotri, M.C. Yildiz, S. Ono, C.-K. Koh, P.H. Madden, Recursive bisection based mixed block placement, in *Proceedings of the 2004 International Symposium on Physical Design*, ser. ISPD '04 (Association for Computing Machinery, New York, NY, USA, 2004), pp. 84–89. [Online]. Available: https://doi.org/10.1145/981066.981084
26. J. Zhao, Q. Zhou, Y. Cai, Fast congestion-aware timing-driven placement for island FPGA, in *2009 12th International Symposium on Design and Diagnostics of Electronic Circuits and Systems* (2009), pp. 24–27
27. Intel, Intel Quartus Prime pro edition user guide: design compilation. https://www.intel.com/content/www/us/en/programmable/documentation/zpr1513988353912.html.
28. N. Metropolis, A.W. Rosenbluth, M.N. Rosenbluth, A.H. Teller, E. Teller, Equation of state calculations by fast computing machines **3** (1953)

29. J. Yuan, J. Chen, L. Wang, X. Zhou, Y. Xia, J. Hu, Arbsa: adaptive range-based simulated annealing for FPGA placement. IEEE Trans. Comput.-Aid. Des. Integr. Circ. Syst. **38**(12), 2330–2342 (2019)
30. K.E. Murray, O. Petelin, S. Zhong, J.M. Wang, M. Eldafrawy, J.-P. Legault, E. Sha, A.G. Graham, J. Wu, M.J.P. Walker, H. Zeng, P. Patros, J. Luu, K.B. Kent, V. Betz, Vtr 8: high-performance cad and customizable FPGA architecture modelling. ACM Trans. Reconfigurable Technol. Syst. **13**(2) (2020). [Online]. Available: https://doi.org/10.1145/3388617
31. K. Vorwerk, A. Kennings, J.W. Greene, Improving simulated annealing-based FPGA placement with directed moves. IEEE Trans. Comput.-Aid. Des. Integr. Circ. Syst. **28**(2), 179–192 (2009)
32. M.A. Elgammal, K.E. Murray, V. Betz, Rlplace: using reinforcement learning and smart perturbations to optimize FPGA placement. IEEE Trans. Comput.-Aid. Des. Integr. Circ. Syst. **41**(8), 2532–2545 (2022)
33. A. Ludwin, V. Betz, Efficient and deterministic parallel placement for FPGAs **16**(3) (2011). [Online]. Available: https://doi.org/10.1145/1970353.1970355
34. T. Kong, A novel net weighting algorithm for timing-driven placement, in *IEEE/ACM International Conference on Computer Aided Design, 2002. ICCAD 2002* (2002), pp. 172–176
35. H. Ren, D. Pan, D. Kung, Sensitivity guided net weighting for placement-driven synthesis. IEEE Trans. Comput.-Aid. Des. Integr. Circ. Syst. **24**(5), 711–721 (2005)
36. A. Marquardt, V. Betz, J. Rose, Timing-driven placement for FPGAs in *FPGA '00* (2000)
37. K. Eguro, S. Hauck, Enhancing timing-driven FPGA placement for pipelined netlists, in *2008 45th ACM/IEEE Design Automation Conference* (2008), pp. 34–37
38. C. Guth, V. Livramento, R. Netto, R. Fonseca, J.L. Güntzel, L. Santos, Timing-driven placement based on dynamic net-weighting for efficient slack histogram compression, in *Proceedings of the 2015 Symposium on International Symposium on Physical Design*, ser. ISPD '15 (Association for Computing Machinery, New York, NY, USA, 2015), pp. 141–148. [Online]. Available: https://doi.org/10.1145/2717764.2717766
39. W. Swartz, C. Sechen, Timing driven placement for large standard cell circuits, in *Proceedings of the 32nd Annual ACM/IEEE Design Automation Conference*, ser. DAC '95 (Association for Computing Machinery, New York, NY, USA, 1995), pp. 211–215. [Online]. Available: https://doi.org/10.1145/217474.217531
40. A. Chowdhary, K. Rajagopal, S. Venkatesan, T. Cao, V. Tiourin, Y. Parasuram, B. Halpin, How accurately can we model timing in a placement engine? in *Proceedings 42nd Design Automation Conference* (2005) pp. 801–806
41. T. Luo, D. Newmark, D.Z. Pan, A new lp based incremental timing driven placement for high performance designs, in *2006 43rd ACM/IEEE Design Automation Conference* (2006), pp. 1115–1120
42. N. Viswanathan, G.-J. Nam, J. A. Roy, Z. Li, C. J. Alpert, S. Ramji, C. Chu, Itop: integrating timing optimization within placement, in *Proceedings of the 19th International Symposium on Physical Design*, ser. ISPD '10 (Association for Computing Machinery, New York, NY, USA, 2010), pp. 83–90. [Online]. Available: https://doi.org/10.1145/1735023.1735048
43. J. Lamoureux, S.J.E. Wilton, On the interaction between power-aware FPGA cad algorithms, in *ICCAD-2003. International Conference on Computer Aided Design (IEEE Cat. No.03CH37486)* (2003) pp. 701–708
44. W.-J. Sun, C. Sechen, A parallel standard cell placement algorithm. IEEE Trans. Comput.-Aid. Des. Integr. Circ. Syst. **16**(11), 1342–1357 (1997)
45. S. Birk, J.G. Steffan, J.H. Anderson, Parallelizing FPGA placement using transactional memory, in *2010 International Conference on Field-Programmable Technology* (2010), pp. 61–69
46. J.B. Goeders, G.G. Lemieux, S.J. Wilton, Deterministic timing-driven parallel placement by simulated annealing using half-box window decomposition, in *2011 International Conference on Reconfigurable Computing and FPGAs* (2011), pp. 41–48
47. B. Huang, H. Zhang, Application of multi-core parallel computing in FPGA placement, in *2013 2nd International Symposium on Instrumentation and Measurement, Sensor Network and Automation (IMSNA)* (2013), pp. 884–889

48. M. An, J.G. Steffan, V. Betz, Speeding up FPGA placement: parallel algorithms and methods, in *2014 IEEE 22nd Annual International Symposium on Field-Programmable Custom Computing Machines* (2014), pp. 178–185
49. C. Fobel, G. Grewal, D. Stacey, A scalable, serially-equivalent, high-quality parallel placement methodology suitable for modern multicore and FPU architectures, in *2014 24th International Conference on Field Programmable Logic and Applications (FPL)* (2014), pp. 1–8
50. A. Al-Kawam, H.M. Harmanani, A parallel GPU implementation of the timber wolf placement algorithm, in *2015 12th International Conference on Information Technology - New Generations* (2015), pp. 792–795
51. M.G. Wrighton, A.M. DeHon, Hardware-assisted simulated annealing with application for fast FPGA placement, in *Proceedings of the 2003 ACM/SIGDA Eleventh International Symposium on Field Programmable Gate Arrays*, ser. FPGA '03 (Association for Computing Machinery, New York, NY, USA, 2003), pp. 33–42. [Online]. Available: https://doi.org/10.1145/611817.611824
52. C. Tian, L. Chen, Y. Wang, S. Wang, J. Zhou, Y. Zhang, G. Li, Improving simulated annealing algorithm for FPGA placement based on reinforcement learning, in *2022 IEEE 10th Joint International Information Technology and Artificial Intelligence Conference (ITAIC)*, vol. 10 (2022), pp. 1912–1919
53. W. Li, S. Dhar, D.Z. Pan, Utplacef: a routability-driven FPGA placer with physical and congestion aware packing. IEEE Trans. Comput-Aid. Des. Integr. Circ. Syst. **37**(4), 869–882 (2018)
54. C.-W. Pui, G. Chen, W.-K. Chow, K.-C. Lam, J. Kuang, P. Tu, H. Zhang, E.F.Y. Young, B. Yu, RippleFPGA: a routability-driven placement for large-scale heterogeneous FPGAs, in *2016 IEEE/ACM International Conference on Computer-Aided Design (ICCAD)* (2016), pp. 1–8
55. R. Pattison, Z. Abuowaimer, S. Areibi, G. Gráwal, A. Vannelli, Gplace: a congestion-aware placement tool for ultrascale FPGAs, in *2016 IEEE/ACM International Conference on Computer-Aided Design (ICCAD)* (2016), pp. 1–7
56. Y. Meng, W. Li, Y. Lin, D.Z. Pan, "elf place: electrostatics-based placement for large-scale heterogeneous FPGAs. IEEE Trans. Comput.-Aid. Des. Integr. Circ. Syst. **41**(1), 155–168 (2022)
57. M. Gort, J.H. Anderson, Analytical placement for heterogeneous FPGAs, in *22nd International Conference on Field Programmable Logic and Applications (FPL)* (2012) pp. 143–150
58. D. Vercruyce, E. Vansteenkiste, D. Stroobandt, Liquid: high quality scalable placement for large heterogeneous FPGAs, in *2017 International Conference on Field Programmable Technology (ICFPT)* (2017), pp. 17–24
59. Z. Abuowaimer, D. Maarouf, T. Martin, J. Foxcroft, G. Gréwal, S. Areibi, A. Vannelli, Gplace3.0: routability-driven analytic placer for ultrascale FPGA architectures. ACM Trans. Des. Autom. Electron. Syst. **23**(5) (2018). [Online]. Available: https://doi.org/10.1145/3233244
60. W. Li, D.Z. Pan, A new paradigm for FPGA placement without explicit packing. IEEE Trans. Comput.-Aid. Des. Integr. Circ. Syst. **38**(11), 2113–2126 (2019)
61. T. Lin, C. Chu, Polar 2.0: an effective routability-driven placer, in *2014 51st ACM/EDAC/IEEE Design Automation Conference (DAC)* (2014), pp. 1–6
62. R. Pattison, C. Fobel, G. Grewal, S. Areibi, Scalable analytic placement for FPGA on GPGPU, in *2015 International Conference on ReConFigurable Computing and FPGAs (ReConFig)* (2015), pp. 1–6
63. C.-X. Lin, M.D.F. Wong, Accelerate analytical placement with GPU: a generic approach, in *2018 Design, Automation and Test in Europe Conference and Exhibition (DATE)* (2018), pp. 1345–1350
64. S. Dhar, L. Singhal, M.A. Iyer, D.Z. Pan, FPGA-accelerated spreading for global placement, *2019 IEEE High Performance Extreme Computing Conference (HPEC)* (2019), pp. 1–7
65. A. Al-Hyari, A. Shamli, T. Martin, S. Areibi, G. Grewal, An adaptive analytic FPGA placement framework based on deep-learning, in *2020 ACM/IEEE 2nd Workshop on Machine Learning for CAD (MLCAD)* (2020), pp. 3–8

66. H. Wang, X. Tong, C. Ma, R. Shi, J. Chen, K. Wang, J. Yu, Y.-W. Chang, CNN-inspired analytical global placement for large-scale heterogeneous FPGAs, in *Proceedings of the 59th ACM/IEEE Design Automation Conference*, ser. DAC '22 (Association for Computing Machinery, New York, NY, USA, 2022), pp. 637–642
67. P. Esmaeili, T. Martin, S. Areibi, G. Grewal, Guiding FPGA detailed placement via reinforcement learning, in *2022 IFIP/IEEE 30th International Conference on Very Large Scale Integration (VLSI-SoC)* (2022), pp. 1–6
68. T. Liang, G. Chen, J. Zhao, S. Sinha, W. Zhang, AMF-placer 2.0: open source timing-driven analytical mixed-size placer for large-scale heterogeneous FPGA (2022)
69. K.E. Murray, S. Zhong, V. Betz, Air: A fast but lazy timing-driven FPGA router, in *2020 25th Asia and South Pacific Design Automation Conference (ASP-DAC)* (2020), pp. 338–344
70. V. Developers, Routing resource graph-vtr. https://docs.verilogtorouting.org/en/latest/api/vpr/rr_graph/
71. L. McMurchie, C. Ebeling, Pathfinder: a negotiation-based performance-driven router for FPGAs, in *Third International ACM Symposium on Field-Programmable Gate Arrays* (1995), pp. 111–117
72. R. Wood, R. Rutenbar, FPGA routing and routability estimation via Boolean satisfiability. IEEE Trans. Very Large Scale Integr. (VLSI) Syst. **6**(2), 222–231 (1998)
73. G.-J. Nam, F. Aloul, K. Sakallah, R. Rutenbar, A comparative study of two Boolean formulations of FPGA detailed routing constraints. IEEE Trans. Comput. **53**(6), 688–696 (2004)
74. S. Mukherjee, S. Roy, Sat based multi pin net detailed routing for FPGA, in *2010 International Symposium on Electronic System Design* (2010), pp. 141–146
75. V. Chopra, C. Deptt, P. India, A. Singh, P.P. India, Ant colony optimization approach for solving FPGA routing with minimum channel width (2011)
76. H. Fraisse, A. Joshi, D. Gaitonde, A. Kaviani, Boolean satisfiability-based routing and its application to Xilinx ultrascale clock network, in *Proceedings of the 2016 ACM/SIGDA International Symposium on Field-Programmable Gate Arrays*, ser. FPGA '16 (Association for Computing Machinery, New York, NY, USA, 2016), pp. 74–79. [Online]. Available: https://doi.org/10.1145/2847263.2847342
77. S. Boshra, H. Abbas, A. Darwish, I. Talkhan, Performance and routability improvements for routability-driven FPGA routers, in *2006 IEEE International Symposium on Circuits and Systems (ISCAS)* (2006) pp. 4
78. D. Vercruyce, E. Vansteenkiste, D. Stroobandt, Croute: a fast high-quality timing-driven connection-based FPGA router, in *2019 IEEE 27th Annual International Symposium on Field-Programmable Custom Computing Machines (FCCM)* (2019), pp. 53–60
79. Y. Zhou, P. Maidee, C. Lavin, A. Kaviani, D. Stroobandt, RWRoute: an open-source timing-driven router for commercial FPGAs. ACM Trans. Reconfigurable Technol. Syst. **15**(1) (2021). [Online]. Available: https://doi.org/10.1145/3491236
80. C.H. Hoo, Y. Ha, A. Kumar, A directional coarse-grained power gated FPGA switch box and power gating aware routing algorithm, in *2013 23rd International Conference on Field programmable Logic and Applications* (2013), pp. 1–4
81. C. H. Hoo, A. Kumar, Y. Ha, Paralar: A parallel FPGA router based on Lagrangian relaxation, in *2015 25th International Conference on Field Programmable Logic and Applications (FPL)* (2015), pp. 1–6.
82. D. Wang, Z. Duan, C. Tian, B. Huang, N. Zhang, Parra: a shared memory parallel FPGA router using hybrid partitioning approach. IEEE Trans Comput-Aided Des Integrated Circ Syst **39**(4), 830–842 (2020)
83. M. Shen, G. Luo, N. Xiao, Coarse-grained parallel routing with recursive partitioning for FPGAs. IEEE. Trans. Parallel Distrib. Syst. **32**(4), 884–899 (2021)
84. Y. Moctar, M. Stojilović, P. Brisk, Deterministic parallel routing for FPGAs based on Galois parallel execution model, in *2018 28th International Conference on Field Programmable Logic and Applications (FPL)* (2018) pp. 21–214.
85. J.F. Croix, S.P. Khatri, Introduction to GPU programming for EDA, in *2009 IEEE/ACM International Conference on Computer-Aided Design—Digest of Technical Papers* (2009) pp. 276–280

86. M. Shen, G. Luo, Corolla: GPU-accelerated FPGA routing based on subgraph dynamic expansion," in *Proceedings of the 2017 ACM/SIGDA International Symposium on Field-Programmable Gate Arrays*, ser. FPGA '17 (Association for Computing Machinery, New York, NY, USA, 2017) pp. 105–114. [Online]. Available: https://doi.org/10.1145/3020078.3021732
87. D. Korolija, M. Stojilović, FPGA-assisted deterministic routing for FPGAs, in *2019 IEEE International Parallel and Distributed Processing Symposium Workshops (IPDPSW)* (2019), pp. 155–162
88. U. Farooq, N. Ul Hasan, I. Baig, M. Zghaibeh, Efficient FPGA routing using reinforcement learning, in *2021 12th International Conference on Information and Communication Systems (ICICS)* (2021), pp. 106–111
89. T. Martin, C. Barnes, G. Grewal, S. Areibi, Integrating machine-learning probes into the VTR FPGA design flow, in *2022 35th SBC/SBMicro/IEEE/ACM Symposium on Integrated Circuits and Systems Design (SBCCI)* (2022), pp. 1–6

Chapter 11
Bitstream Configuration

Abstract This chapter will introduce the final step of FPGA application design EDA–bitstream configuration, including bitstream generation, compression, encryption, and programming. Bitstream is the bottom level machine code that actually makes FPGA work, after all the efforts made by previous EDA engines are assembled in this image by the bitstream generator, it will be downloaded into the FPGA by the bitstream programmer. However, modern FPGA technology also introduced challenges, such as the excessive growth of the file size and stealthy attack threats against the programming system. Bitstream compression and encryption is thereby rise to the challenges.

11.1 Bitsream Generation

11.1.1 Overview

In FPGA application design flow, after physical implementation is complete, all configuration information (in the design checkpoint) then has been determined. Based on this information, the bitstream generation process is ready to launch.

The bitstream configuration system of commercial tools (such as AMD Vivado and Intel Quartus) has been kept as secret because it is highly related to proprietary hardware architecture. In academic world, studies on bitstream focused on these aspects (Table 11.1): reverse engineering, manipulation interface, and conventional generation.

1. Reverse Engineering
   Due to lack of commercial FPGA's bitstream details, reverse engineering has been intensively studied by academic researchers. Strictly speaking, it is not about "generate the bitstream", but to derive the original design from bitstream. Mature commercial FPGA devices are the most ideal targets for such research.
   Benefit from their popularity, AMD/Xilinx devices are the most frequently addressed. Debit [1], BIL [2], DAT [3], Bit2NCD [4], BRET [5], BRT/NRT [6], X-Ray/U-Ray [7–9] are the representative works in this category.

K. Tu et al., *FPGA EDA*, https://doi.org/10.1007/978-981-99-7755-0_11

Lattice devices are another hot spot because of their simplicity. Icestorm/Trellis [10, 11] projects are well known among these type of works.
Besides, Microchip devices related attempts [12] are emerging as well.
Although reverse engineering has been called the "dark side" of semiconductor industry [13], studies in this field do have their unique value, that is, revealing possible security threats and avoiding potential attacks [14, 15].

2. Manipulation Interface
   Running the full EDA flow and get the final bitstream is time consuming and sometimes unnecessary. In order to modify the bitstream quickly and precisely without invoking the previous intricate EDA processes such as place and route, many research works (or even the vendor themselves) offer low level manipulation interfaces (such as JHDLBits [16], JBits [17], Abits [18], BitMan [19], MaNaBit [20], RapidSmith [21], RapidWright [22]). These APIs give the user a greater freedom of direct interact with the bitstream without knowing too much bit level information of the device.
3. Conventional Generation
   Having known all the details of FPGA device library mentioned in Sect. 2.2 (can also be derived by reverse engineering), the conventional bitstream generation process extract information from design database/checkpoint and output the bitstream under the specified configuration protocol. The bitstream generation efficiency (time) is the main concern for methods in this field. The bitstream generation time can be divided into three parts: time for device database loading, time for bitstream configuration, and time for bitstream file writing. Among them, time for bitstream configuration can be various depends on the method it takes, and the other two times are relatively fixed.

11.1.2 Mode-Based Technique

Mode-based bitstream generation methods are the most commonly used in industry. The work flow is shown in Fig. 11.1.

The flow begins with the given application design database, which contains all the Programmable Point (PP) information that needs to be configured. After all the PP information is extracted, configuration mode matching is performed to decide each configuration bit's logical address, then the address mapping phase finds out the correlation of logical to physical address of each configuration bit, at last, the final bitstream is written out based on the physical addresses of configuration bits.

1. Configuration mode matching
   Having discussed in Sect. 2.2.3, PP is defined to be the basic configuration element, it may include one or multiple atoms/primitives' configure information. In bitstream generation guidance model, all PPs are pre-defined by illustrating all their possible configuration modes, each mode carrying bits' logical address that need to be configured. All effort made by previous EDA steps, to some extent, is a process that gives every PP a certain mode in the design database.

Table 11.1 Bitstream related representative research works

Work	Type	Targeted FPGA	Year
JBits [17]	Manipulation Interface	AMD/Xilinx XC4000 and Virtex	2001
JHDLBits [16]	Manipulation Interface	AMD/Xilinx Virtex II	2004
Abits [18]	Manipulation Interface	Microchip/Atmel FPSLIC	2007
Debit [1]	Reverse Engineering	AMD/Xilinx Spartan-3/Virtex-2,3,4,5	2008
BIL [2]	Reverse Engineering	AMD/Xilinx Virtex-5	2012
DAT [3]	Reverse Engineering	AMD/Xilinx Spartan-3	2013
Bit2NCD [4]	Reverse Engineering	AMD/Xilinx Spartan-3,3E/Virtex-2,4,5	2013
[23]	Conventional Generation	Any architecture	2013
Icestorm [10]	Reverse Engineering	Lattice iCE40	2015
RapidSmith [21]	Manipulation Interface	AMD/Xilinx	2015
BitMan [19]	Manipulation Interface	AMD/Xilinx 6, 7, US, US+	2017
BRET [5]	Reverse Engineering	AMD/Xilinx Virtex-5	2018
RapidWright [22]	Manipulation Interface	AMD/Xilinx	2018
BRT/NRT [6]	Reverse Engineering	AMD/Xilinx Spartan-6	2019
Trellis [11]	Reverse Engineering	Lattice ECP5	2019
X-Ray [7, 8]	Reverse Engineering	AMD/Xilinx 7	2020
U-Ray [9]	Reverse Engineering	AMD/Xilinx US, US+	2020
[12]	Reverse Engineering	Microchip/Microsemi ProASIC3	2021
MaNaBit [20]	Manipulation Interface	AMD/Xilinx 7, US+	2022

2. Bit address mapping
   Once the configuration mode is fully matched, logical address of each configuration bit is identified. Bit address mapping then derives bit's physical address from its logical address in the device, under the help of device configuration bit structure model Fig. 2.21 mentioned in Sect. 2.3. This mapping process is commonly hierarchical, that is, figure out the local physical address in the first place and then adjust it with offsets to recursively achieve the global physical address.

The future research directions of mode-based bitstream generation technique includes:

1. Reducing generation time
   Software or hardware acceleration/parallel schemes can be applied to bitstream generator to shorten running time, just like the situation of placer.
2. Improving guidance model
   More simple and efficient guidance model can help to shrink the memory footprint.

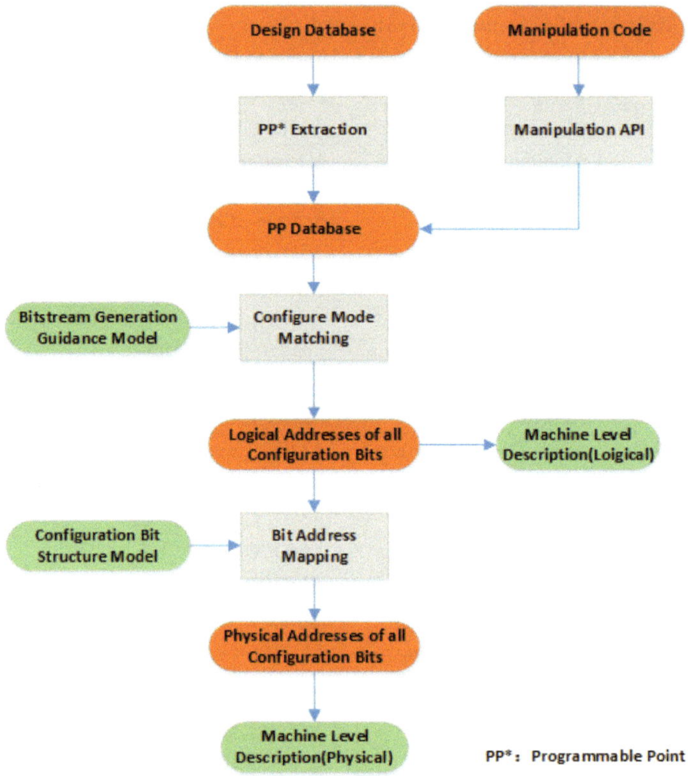

Fig. 11.1 FPGA bitstream generation conventional flow

11.2 Bitsream Compression

11.2.1 Overview

The original bitstream file size based on application design CBS model (see Sect. 3.2) is related to the application design size. Thanks to the fast shrinking of technology nodes, the resource capacity of FPGA device has been bloomed dramatically and therefore can accommodate more intricate application designs. However, the increasing of bitstream file size could possibly bring memory footprint and communication bandwidth problems, and consequently becomes a drawback that can not be neglected. Bitstream compression and decompression is the most straightforward way to overcome this obstacle–compression outside the FPGA by software and decompression inside the FPGA by hardware (Fig. 11.2). In this section, we will discuss several renowned bitstream compression methods.

The efficiency of bitstream compression is measured by Compression Ratio (CR) and Compression Time (CT). CR is the ratio between the compressed size and the

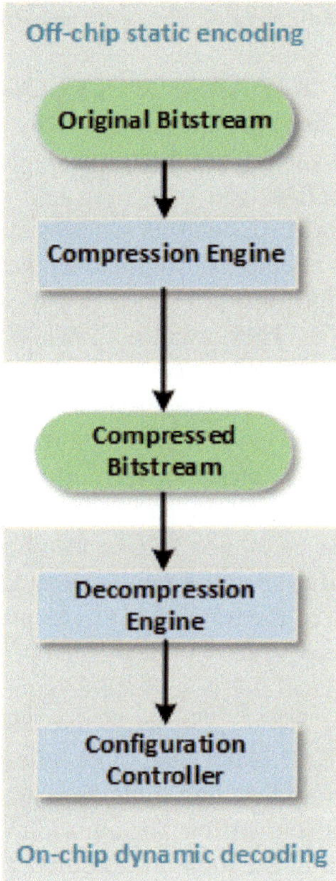

Fig. 11.2 FPGA bitstream compression and decompression flow

original size, CT is the time spent for the compression process. The smaller CR and CT indicates more effective compression. CR and CT are the most important metrics that compression algorithms must concern.

Entropy is a concept borrowed from information theory, it measures the disorder of the target information data. Entropy encoding methods consider the entropy of the bitstream and therefore can often achieve a lower CR, yet the non-entropy encoding methods don't, so they can achieve lower CT.

11.2.2 Non-entropy Encoding

1. Run-Length Encoding (RLE)
   Run-Length Encoding (RLE) is a typical non-entropy coding method, and it's one of the most deceptively simple and powerful encoding techniques. The principle is to replace runs of same data by the count and the data value only once. When

input include digits, a translation must be done to avoid conflict between data and counts.
For example, a string of "AAAAAFDDBBCCCCCCC" can be compressed to "5A1F2D2B7C", that is, from 17 characters to 10.
Applications of RLE in bitstream compression are deeply studied [24–28].

2. Lempel-Ziv Encoding (LZE)
The Lempel-Ziv Encoding (LZE) and its variants are so classic, that in the past decades, there hasn't been another algorithm to replace them. LZ77 [29] and LZ78 [30] are the first LZE algorithms published in papers by Abraham Lempel and Jacob Ziv in 1977 and 1978. They are also known as LZ1 and LZ2, respectively. LZE algorithms work by defining a fixed-size dictionary to hold bytes from an input bitstream, and then referring to the dictionary when compressing the remainder of the input source to find existing patterns. If a pattern in the input source is already in the dictionary, this pattern is replaced with a reference to the position in the dictionary and the length of the pattern.
Take LZ77 as an example, assume a 6-byte search buffer (dictionary) size and an 3-byte look-ahead buffer size, the first 6 bytes from the input bitstream are loaded into the search buffer, the following 3 bytes are loaded into the look-ahead buffer. Then search for a sequence in the search buffer that begins with the byte in look-ahead buffer position 0 ("B"). Such a sequence of three bytes starts at search buffer position 2 ("BBC"). So these three bytes can be replaced by an (Offset, Length), that is (2, 3). After that, three bytes from the look-ahead buffer are shifted into the search buffer, and three new bytes from the input bitstream are shifted into the look-ahead buffer. The algorithm keeps doing this iteratively until all the bytes shift out of the look-ahead buffer (Fig. 11.3) .

Non-entropy encoding methods are currently most adopted by FPGA industry because they are mature and relatively easier to implement. LZ77/LZ78/LZSS/LZW are the representative variants of LZE and has been proudly used for commercial FPGAs [31–35].

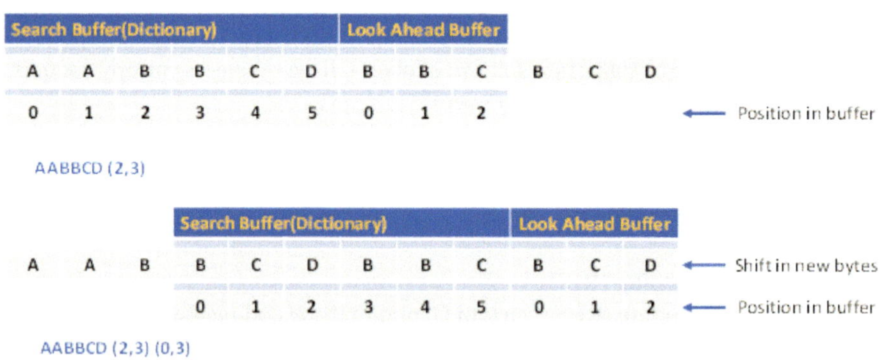

Fig. 11.3 LZ77 compression example

11.2.3 Entropy Encoding

1. Huffman Encoding
   Huffman encoding is a typical entropy encoding method developed by David Huffman in 1952 [36]. This method is based on the building a full binary tree for the different symbols that are in the original file after calculating the probability of each symbol and put in descending order [37, 38].
   For example, there is an initial string of "BAAADDDCCCACACA", Huffman encoding first calculate the frequency of each character in the string and sort the characters in increasing order of the frequency (Fig. 11.4).
   To build the Huffman tree, make each unique character as a leaf node. Create an node $N0$, assigning the minimum frequency to the left child of $N0$ and the second minimum frequency to the right child of $N0$, setting the frequency value of the $N0$ as the sum of the above two minimum frequencies. After that, repeat the same actions for the two nodes that currently have the lowest frequencies ($N0$ and C) until all nodes are involved. For each non-leaf node, assign 0 to the left edge and 1 to the right edge (Fig. 11.5). After encoding, A, B, C, D would be 0, 100, 11, 101.

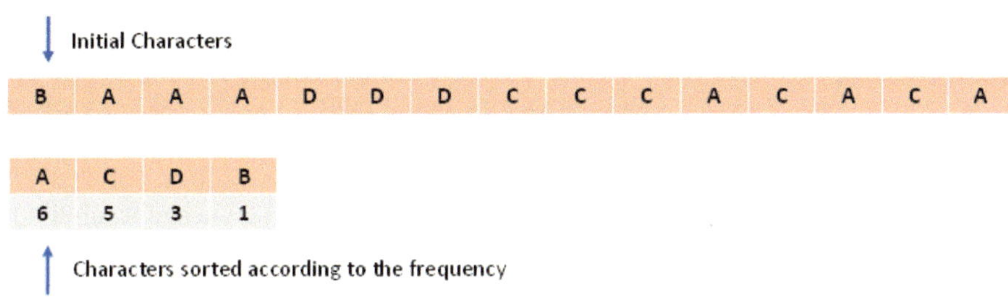

Fig. 11.4 Huffman compression example—ordering characters by frequency

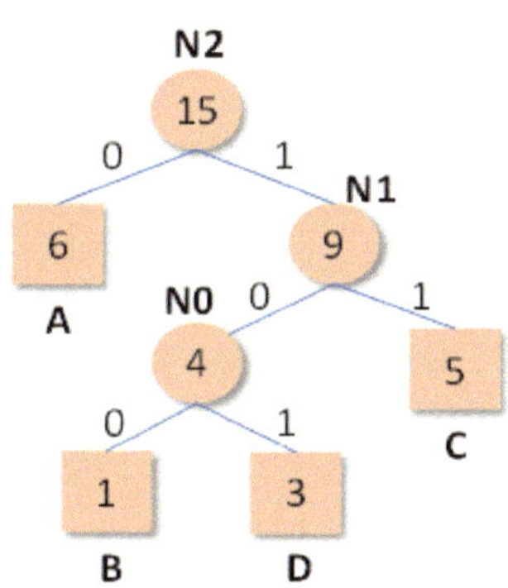

Fig. 11.5 Huffman compression example—Huffman tree

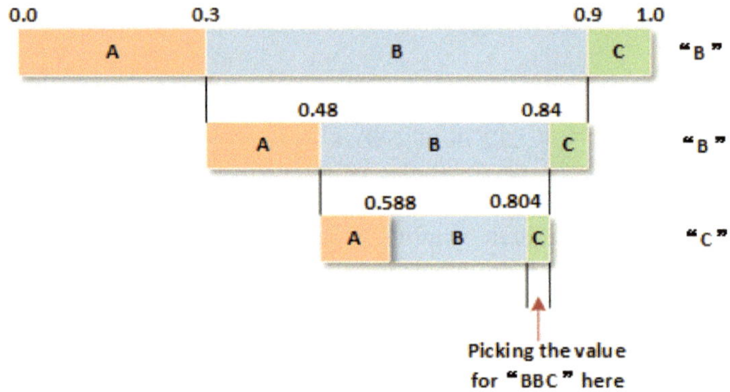

Fig. 11.6 Arithmetic compression example

2. Arithmetic Encoding
   Arithmetic encoding considers the sequence characteristic and estimates the conditional probability of each bit. It allows storing symbols using a fractional number of bits based on the probability of occurrence [39, 40].
   For example, a string with three kinds of characters–A, B, and C, with probabilities of 0.3, 0.6, and 0.1, respectively. We assign them ranges in the interval [0, 1) according to their probability. If the target string needs encoding is "BBC", as seen in (Fig. 11.6), the first character from the input stream is "B", according to the probability, the interval of "B" is [0.3, 0.9), then we subdivide the number space between the values of 0.3 and 0.9 according to probabilities, which gives us a new range of intervals. The second character is also "B", we again subdivide the new interval of "B" according to the probability, and make a new range of intervals. The third character is "C", we finally pick a value (e.g., 0.82) from the newest interval of "C".

   The future research directions of entropy encoding for FPGA bitstream includes:

1. Characteristics comparison
   When dealing with bitstream, which has a highly specialized structure, encoding characteristics (such as CR, CT, and memory footprint) comparison among entropy and non-entropy methods can be surveyed.
2. AI-assist compression
   Compressing data with the help of AI technology is attracted by the academic community to further improve CR. For example, neural network-based probability predictor can be used to accurate estimate the probability distribution of each bit during arithmetic encoding process [41, 42].

11.3 Bitsream Encryption

11.3.1 Overview

FPGAs security issues have been with them since their inception. Bitstream is the only vessel that finally carry the configuration information, and naturally becomes the primal target by the attackers [43] (Fig. 11.7).

FPGA bitstream life cycle usually has five stages:

1. Bitstream-Generation
   The stage that the bitstream is being generated by EDA tools.
2. Bitstream-At-Rest
   The stage that the bitstream has been generated and is stored in a non-volatile memory that is not currently configuring the FPGA.
3. Bitstream-Loading
   The stage that the bitstream is being loaded into the FPGA configuration memory.
4. Bitstream-Running
   The stage that the bitstream has been loaded into the FPGA configuration memory and the FPGA is operating.
5. Bitstream-End-Of-Life
   The stage that the bitstream has been decommissioned.

To meet various bitstream security challenges at each stage, modern FPGAs rely on both encryption and authentication [44]. Encryption provides the basic design security to protect the design from copying or reverse engineering, while authentication provides assurance that the bitstream is the original and intact one created by an authorized user. In industry, decryption are usually done on-chip by hardware, while encryption is responsible for EDA software. Since our book focuses on the EDA side, we emphatically discuss encryption in this section.

The most commonly used bitstream encryption method in industry (both AMD and Intel) is Advanced Encryption Standard (AES) [45].

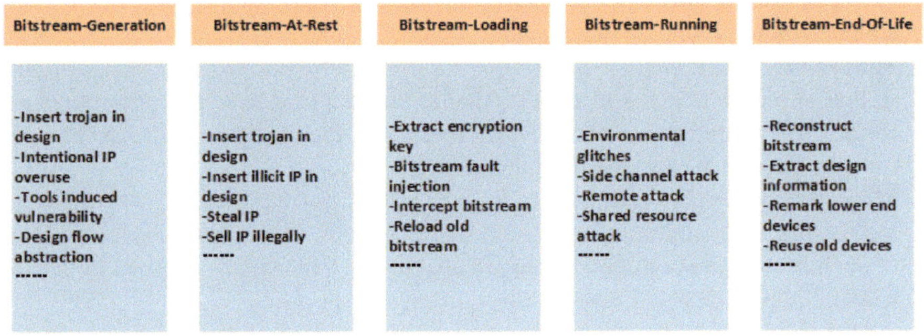

Fig. 11.7 FPGA bitstream security threats [43]

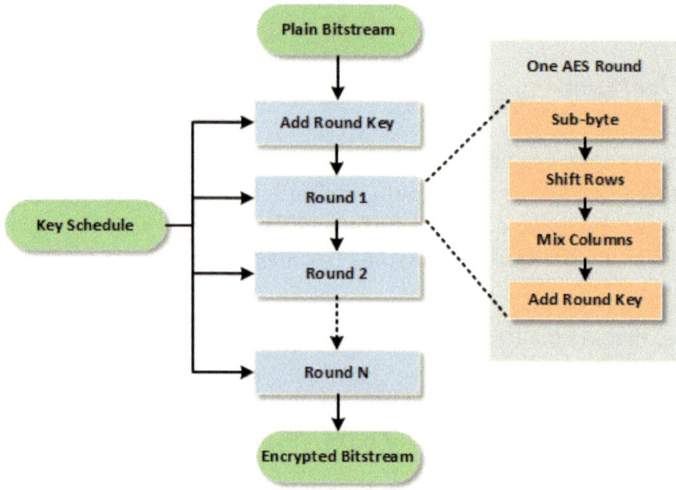

Fig. 11.8 AES algorithm flow for bitstream encryption

11.3.2 AES-Based Technique

Advanced Encryption Standard (AES) is a symmetric encryption algorithm developed by two Belgian cryptographer Joan Daemen and Vincent Rijmen [46]. AES was designed to be efficient in both hardware and software, and supports a block length of 128 bits and key lengths of 128, 192, and 256 bits. Other input, output, and cipher key lengths are not permitted by this standard. AES has 10 rounds for 128-bit keys, 12 rounds for 192-bit keys, and 14 rounds for 256-bit keys.

AES algorithm follows the steps shown in Fig. 11.8.

AES considers each block as a 16 byte (4byte $\times$ 4 = 128 bits) grid and each round has four steps: sub-byte, shift rows, mix columns, and add round key.

1. Sub-byte
   In this step each byte is substituted by another byte in a pre-defined lookup table called substitution box(S-box).
2. Shift rows
   In this step each row is shifted a particular number of times. The first row is not shifted, the second row is shifted once to the left, the third row is shifted twice to the left, the fourth row is shifted thrice to the left.
3. Mix columns
   In this step each column is multiplied with a specific matrix and thus the position of each byte in the column is changed as a result. This step is skipped in the last round.
4. Add round key
   In this step the resultant output of the previous stage is XOR-ed with the corresponding round key. Here, the 16 bytes is not considered as a grid but just as 128 bits of data.

After one cycle, 128 bits of encrypted data is given back as output, and this process is repeated until all the data to be encrypted are treated.

11.4 Bitstream Programming

11.4.1 Overview

Program the bitstream file into the targeted FPGA device is the final ritual of the EDA flow. There are a number of methods to program FPGAs that can be broadly described as parallel or serial, master or slave. In the slave modes (Fig. 11.9), the FPGA is controlling the configuration upon power-up or when triggered by a configuration pin, while in master modes (Fig. 11.10) a non-volatile memory (external or internal) controls the configuration interface. Serial configuration is slower than parallel but uses less signals, therefore leaving more signals to be used for the application itself (Table 11.2).

Joint Test Action Group (JTAG) is a special serial programming mode that can be done via a boundary scan interface (Fig. 11.11). JTAG has TCK, TMS, TDI, and TDO lines for communication, the configuration process can be either master (from non-volatile memory) or slave (from processor).

Bitstream programming (or the programmer) needs co-work of on-chip hardware (downloading circuit) and EDA software, just like bitstream encryption and decryption. For different commercial FPGA structures, downloading circuits are proprietary [47].

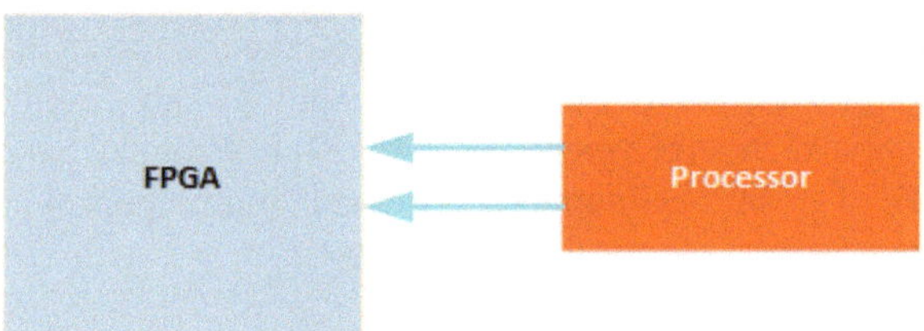

Fig. 11.9 Illustration of FPGA slave mode configuration

Fig. 11.10 Illustration of FPGA master mode configuration (left for external memory, right for internal memory)

Table 11.2 Bitstream configuration mode comparison

Modes	Master or Slave	Serial or Parallel	External device dependency	Width of data bus (in bits)	Relative configuration time
MS	Master	Serial	Memory + Cable	1	Moderate
MP	Master	Parallel	Memory + Cable	>1	Moderate
SS	Slave	Serial	Cable	1	Slow
SP	Slave	Parallel	Cable	>1	Fast

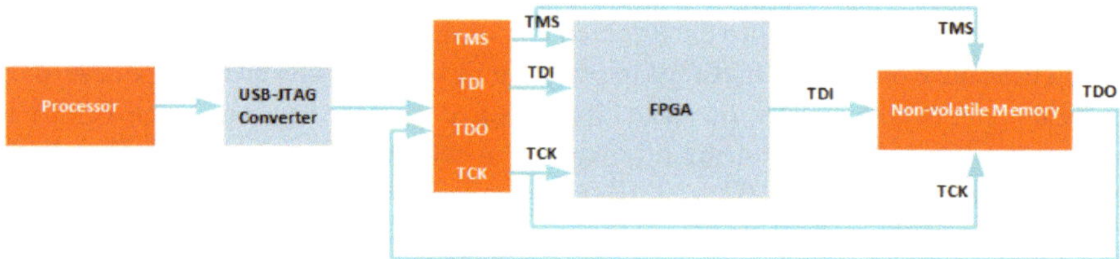

Fig. 11.11 Illustration of FPGA JTAG mode configuration

11.4.2 JTAG-Based Technique

For JTAG (slave) configuration mode, both software and hardware are needed to working together.

On the software side, Open On-Chip Debugger (OpenOCD) is an open-source project that aims to provide debugging, in-system programming, and boundary scan using a debug adapter [48].

On the hardware side, a adapter (commonly USB to JTAG) is a hardware module that provides the right signals for the target to understand. Most of these adapters are based on FT2232H (a chip made from Future Technology Devices International Ltd.) [49], which supports an interface mode known as Multi-Protocol Synchronous Serial Engine (MPSSE). The MPSSE supports a number of complex synchronous-serial operations, some of which are specifically designed for use with JTAG.

References

1. J.B. Note, É. Rannaud, *From the Bitstream to the Netlist. Proceedings of the ACM/SIGDA 16th International Symposium on Field Programmable Gate Arrays (FPGA)* (2008)
2. S.A.F. Benz, H.S.A, Eds., Bil: a tool-chain for bitstream reverse-engineering, in *22nd International Conference on Field Programmable Logic and Applications (FPL)*, (2012)
3. D. Cheremisinov, Design automation tool to generate EDIF and VHDL descriptions of circuit by extraction of FPGA configuration, in *East-West Design and Test Symposium (EWDTS 2013)* (2013), pp. 1–4
4. Z. Ding, Q. Wu, Y. Zhang, Zhu, L. Deriving an ncd file from an FPGA bitstream: methodology, architecture and evaluation. Microproc. Microsyst. **37**(3), 299–312 (2013)

5. J. Yoon, Y. Seo, J. Jang, M. Cho, J. Kim, H. Kim, T. Kwon, A bitstream reverse engineering tool for FPGA hardware trojan detection (2018), 2318–2320
6. T. Zhang, J. Wang, S. Guo, Z. Chen, A comprehensive FPGA reverse engineering tool-chain: From bitstream to RTL code. IEEE Access (2019)
7. F4PGA, Project x-ray https://github.com/f4pga/prjxray (2022)
8. H. Yu, H.-M. Lee, Y. Shin, Y. Kim, FPGA reverse engineering in Vivado design suite based on x-ray project, in *2019 International SoC Design Conference (ISOCC)*, (2019), pp. 239–240
9. F4PGA, Project u-ray. https://github.com/f4pga/prjuray (2022)
10. F4PGA, Project icestorm. https://github.com/f4pga/icestorm (2022)
11. F4PGA, Project trellis. https://github.com/f4pga/prjtrellis (2022)
12. Y. Kim, E.-G. Jung, C. Kim, Bitstream reverse engineering of microsemi's versatile-based FPGAs, in *2021 IEEE Physical Assurance and Inspection of Electronics (PAINE)* (2021) pp. 1–8
13. S. Wallat, M. Fyrbiak, M. Schlögel, C. Paar, A look at the dark side of hardware reverse engineering—a case study, in *2017 IEEE 2nd International Verification and Security Workshop (IVSW)* (2017), pp. 95–100
14. M. Jeong, J. Lee, E. Jung, Y. H. Kim, K. Cho, Extract LUT logics from a downloaded bitstream data in FPGA, in *2018 IEEE International Symposium on Circuits and Systems (ISCAS)* (2018), pp. 1–5
15. M. Ender, A. Moradi, C. Paar, The unpatchable silicon: a full break of the bitstream encryption of xilinx 7-series FPGAs, in *29th USENIX Security Symposium (USENIX Security 20)* (USENIX Association, 2020), pp. 1803–1819. [Online]. Available: https://www.usenix.org/conference/usenixsecurity20/presentation/ender
16. A. Poetter, J. Hunter, C. Patterson, P. Athanas, B. Nelson, N. Steiner, Jhdlbits: the merging of two worlds **3203** (2004), 414–423
17. P. C, G. S. A, (eds.) JBits^{TM} Design abstractions. *The 9th Annual IEEE Symposium on Field-Programmable Custom Computing Machines (FCCM'01)* (2001)
18. A. Megacz, A library and platform for FPGA bitstream manipulation, in *15th Annual IEEE Symposium on Field-Programmable Custom Computing Machines (FCCM 2007)* (2007), pp. 45–54
19. K. Dang Pham, E. Horta, D. Koch, Bitman: a tool and API for FPGA bitstream manipulations, in *Design, Automation and Test in Europe Conference and Exhibition (DATE), 2017* (2017), pp. 894–897
20. N. Charaf, C. Tietz, D. Goehringer, Manabit: a versatile tool for manipulating and analyzing FPGA bitstreams, in *2022 IEEE 30th Annual International Symposium on Field-Programmable Custom Computing Machines (FCCM)*, (2022), pp. 1–1
21. C. Lavin, M. Padilla, J. Lamprecht, P. Lundrigan, B. Nelson, B. Hutchings, Rapidsmith: do-it-yourself cad tools for xilinx FPGAs, in *2011 21st International Conference on Field Programmable Logic and Applications* (2011), pp. 349–355
22. C. Lavin, A. Kaviani, Rapidwright: enabling custom crafted implementations for FPGAs, in *2018 IEEE 26th Annual International Symposium on Field-Programmable Custom Computing Machines (FCCM)* (2018), pp. 133–140
23. R.K. Soni, N. Steiner, M. French, *Open-Source Bitstream Generation* (IEEE Computer Society, USA, 2013) [Online]. Available: https://doi.org/10.1109/FCCM.2013.45
24. J. Vliegen, N. Mentcns, I. Verbauwhede, A single-chip solution for the secure remote configuration of FPGAs using bitstream compression, in *2013 International Conference on Reconfigurable Computing and FPGAs (ReConFig)* (2013), pp. 1–6
25. R. Jia, F. Wang, R. Chen, X.-G. Wang, H.-G. Yang, JTAG-based bitstream compression for FPGA configuration, in *2012 IEEE 11th International Conference on Solid-State and Integrated Circuit Technology*, (2012), pp. 1–3.

26. F. Duhem, F. Muller, P. Lorenzini, Reconfiguration time overhead on field programmable gate arrays: reduction and cost model. Comput. Dig. Tech. IET **6**, 105–113 (2012)
27. A. Abdelhadi, G.G. Lemieux, Configuration bitstream reduction for SRAM-based FPGAs by enumerating LUT input permutations, in *2011 International Conference on Reconfigurable Computing and FPGAs* (2011), pp. 20–26
28. P. Hemnath, V. Prabhu, Compression of FPGA bitstreams using improved RLE algorithm, in *2013 International Conference on Information Communication and Embedded Systems (ICICES)*, (2013), pp. 834–839.
29. J. Ziv, A. Lempel, A universal algorithm for sequential data compression. IEEE Trans. Inf. Theo. **23**(3), 337–343 (1977)
30. J. Ziv, A. Lempel, Compression of individual sequences via variable-rate coding. IEEE Trans. Inf. Theo. **24**(5), 530–536 (1978))
31. Z. Li, S. Hauck, Configuration compression for virtex FPGAs, in *The 9th Annual IEEE Symposium on Field-Programmable Custom Computing Machines (FCCM'01)* (2001), pp. 147–159
32. A. Khu, Xilinx FPGA configuration data compression and decompression (2001)
33. R. Stefan, S.D. Cotofana, Bitstream compression techniques for virtex 4 FPGAs, in *2008 International Conference on Field Programmable Logic and Applications* (2008), pp. 323–328
34. Y. Gao, H. Ye, J. Wang, J. Lai, FPGA bitstream compression and decompression based on lz77 algorithm and bmc technique, in *2015 IEEE 11th International Conference on ASIC (ASICON)* (2015), pp. 1–4
35. R. Iša, J. Matoušek, A novel architecture for lzss compression of configuration bitstreams within FPGA, in *2017 IEEE 20th International Symposium on Design and Diagnostics of Electronic Circuits and Systems (DDECS)* (2017), pp. 171–176
36. D.A. Huffman, A method for the construction of minimum-redundancy codes. Proc. IRE **40**(9), 1098–1101 (1952)
37. M.R. Ashila, N. Atikah, D.R. Ignatius Moses Setiadi, E.H. Rachmawanto, C.A. Sari, Hybrid AES-Huffman coding for secure lossless transmission, in *2019 Fourth International Conference on Informatics and Computing (ICIC)* (2019), pp. 1–5
38. M.E. Hameed, M.M. Ibrahim, N.A. Manap, A.A. Mohammed, A lossless compression and encryption mechanism for remote monitoring of ecg data using huffman coding and cbc-aes. Future Generat. Comput. Syst. **111**, 829–840 (2020). [Online]. Available: https://www.sciencedirect.com/science/article/pii/S0167739X19313950
39. J. Rissanen, G.G. Langdon, Arithmetic coding. IBM J. Res. Dev. **23**, 149–162, (1979).
40. V. Čeperković, M. Prokin, D. Prokin, Efficient cumulative probability distribution estimation for arithmetic coding, in *2020 9th Mediterranean Conference on Embedded Computing (MECO)* (2020), pp. 1–4
41. M. Goyal, K. Tatwawadi, S. Chandak, I. Ochoa, Deepzip: lossless data compression using recurrent neural networks, in *2019 Data Compression Conference (DCC)* (2019), pp. 575–575
42. J. Wang, Y. Kang, Y. Feng, Y. Li, W. Wu, G. Xing, Lossless compression of bitstream configuration files: towards FPGA cloud, in *2021 IEEE International Conference on Parallel and Distributed Processing with Applications, Big Data and Cloud Computing, Sustainable Computing and Communications, Social Computing and Networking (ISPA/BDCloud/SocialCom/SustainCom)* (2021), pp. 1410–1421
43. A. Duncan, F. Rahman, A. Lukefahr, F. Farahmandi, M. Tehranipoor, FPGA bitstream security: a day in the life, in *2019 IEEE International Test Conference (ITC)* (2019), pp. 1–10
44. AMD/Xilinx, Using encryption and authentication to secure an ultrascale/ultrascale+ FPGA bitstream application note https://docs.xilinx.com/v/u/en-US/xapp1267-encryp-efuse-program (2022)
45. S. Sunkavilli, Z. Zhang, Q. Yu, New security threats on FPGAs: from FPGA design tools perspective in *2021 IEEE Computer Society Annual Symposium on VLSI (ISVLSI)* (2021), pp. 278–283
46. N. Pub, 197: advanced encryption standard (AES), federal information processing standards publication **197**, 441–0311 (2001)

47. J. Wang, L.-g. Chen, and J.-m. Lai, FPGA downloading circuit design and implementation," in *2006 8th International Conference on Solid-State and Integrated Circuit Technology Proceedings* (2006), pp. 1950–1953
48. T.O. Project, Open on-chip debugger. https://openocd.org/ (2023)
49. F.T.D.I. Ltd., Ft2232h dual high speed usb to multipurpose UART/FIFO IC datasheet. http://www.ftdichip.com/Support/Documents/DataSheets/ICs/DS_FT2232H.pdf (2023)

Part VI
Summary and Outlook

Chapter 12
Summary and Outlook

Abstract In this chapter, a brief look back of vanilla FPGA EDA is summarized. Commercial FPGA EDA tools for both chip design and application design is very mature and ML-aided engines are starting to show their superior power against traditional engines. Using high-level languages to unify the programming models across computing platforms is continuously evolving, mainly by the "big two" vendors. On the other hand, using mostly open-sourced EDA tools to democratize vanilla FPGA technology across industry and academia also draws a lot of attention. At the end, future works of this book is presented to make a more comprehensive coverage of FPGA EDA knowledge domain.

12.1 FPGA EDA's Crossroads

There are four top performance art awards in United States of America–Emmy, Grammy, Oscar, and Tony. Respectively, these awards honor outstanding achievements in television, recording, film, and theater. People who have won all four of those awards is given a designation of "EGOT" (Fig. 12.1). Musical theater is the only "live" show among the four types of art (half by the producer, half by the actors' "live" performance), so that the comprehensive capability requirements for actors is probably the highest of all. In the computing chip world, there are also four mainstream architecture: Scalar, Vector, Matrix, and Spatial, typically represented by CPU, GPU, TPU, and FPGA. Every outstanding chip company is eager to become the "EGOT" in the chip industry—collecting all these four technologies. FPGA is the only "field" programmable hardware among the four types of architecture (half by the producer, half by the users' "field" programming), which gives this technology a high entry barrier of learning. Besides, FPGA can be programmed to simulate every other computing architecture, just like an excellent musical artist usually can be easily fitted in any other form of performance.

As the "big two" (Xilinx and Altera) were aquired by AMD and Intel respectively, vanilla FPGA has been reduced to an element serving mighty hybrid computing complex such as Versal or Agilex series chips. Vanilla FPGA EDA (for chip design—from transistor to layout; for application design—from HDL to bitstream) has come

K. Tu et al., *FPGA EDA*, https://doi.org/10.1007/978-981-99-7755-0_12

Fig. 12.1 Four most important rewards of performance art in the USA. *Source* Sporcle

to a crossroads. At this point, industry and academia both turned their focuses mainly into three directions:

1. Keep optimizing vanilla FPGA EDA tools by accelerators or AI technologies.
2. Create new FPGA EDA tools by increasing the level of abstraction that the compiler can understand.
3. Make vanilla FPGA EDA tools open source.

All of these technical routes are serving the same ultimate purpose: democratize the FPGA technology.

At the moment this book is writing, most of the published works mentioned above concentrate at the FPGA application design EDA side. To further improve the quality and speed, AI-aided EDA engines has been applied to AMD's Vivado. To increase the level of abstraction that the compiler can understand, AMD's Vitis unifies the design methodologies and programming models across different computing platforms by using Tensorflow/Caffe/C++/Python etc., whilst Intel's oneAPI tries to further integrate the programming languages into DPC++, a new programming model across all computing platforms. To make tools open source, VTR, Yosys, and many other pioneer masterpieces have been very popular in the learning community and together they have forged the cornerstone of FPGA academic research.

12.2 Our Book's Future Works

A series of new chapters will be added in the next edition to enrich the knowledge spectrum.

For [Part. IV]—FPGA Chip Design EDA, one new chapter will be added.

1. Full-custom EDA
   Introducing full-custom EDA technologies for FPGA chip design studied in academia.

For [Part. V]—FPGA Application Design EDA, three extra chapters will be enrolled.

1. Engine Fusion
   In previous edition, every single EDA engine has been orderly discussed, however, in modern FPGA application design EDA, boundaries across conventional engines are blurred, these engines are often fused together to improve quality-of-results (QoR) or enable new functionalities in practical situations.
2. Generic GUI Framework
   A generic open-source Graphic User Interface (GUI) framework is presented, it follows the classic EDA tool design pattern of Tcl-driven QT GUIs. It offers a collection of widgets and allow users to create their own GUI without having to reinvent the wheel.
3. Test Benchmarks
   The FPGA community relies heavily on standard and fair benchmarks to evaluate their hardware and software solutions. In this chapter, a set of up-to-date benchmark suites are introduced and analyzed to ensure the effectiveness and efficiency of the test activity.